The Preface of Volume 1

Xu Feng

When I wrote down these words for the Volume 1, the day was cloudy with a little cold. But you know, I'm super happy now for every day and for every moment in my life.

The Volume 1 is a tribute to the twin theorems: Anhua-Diophantus Theorem and Beal Theorem. They make me feel so great that you have never known in my heart the twin theorems what they mean.

As you know, each volume is different. The Volume 2 is of the equations of new thing. The Volume 3 is of the complex numbers. The Volume 4 is of the combinations. And more others…

I still remember a funny thing: Everything is China's history. But when you say the truth of China's history, the Chinese should be crying. Why? The Chinese don't like any truth of China's history and of themselves. They just indulge in the fucking GDP for keeping on their fragile faith.

Today, for the fucking GDP, the Chinese are high in anywhere. They say: China and the Chinese rise in the world. But in fact, it is only a joke. China and the Chinese bring nothing to the world. So, don't tell me: Made in China. Of course, you may say that: China and the Chinese offer the cheap labor.

The other cruel truth is that: lots of Chinese study and teach at universities anywhere, they are still hopeless. Sure, they can get the good result of the examination. But so what? They have no any abilities to manage any country anywhere. They are always alone with the poor helplessness to face the world so hard. No ones like them. For the good result of the examination, they call themselves: an excellent race. Haha, how really poor the Chinese are.

And what did I say, Guangdongen? I remind you Guangdongen again: Don't need any Chinese anytime. The Chinese do nothing but making-shit anywhere.

Sure, the Chinese have lots of morbid thoughts. 1) They call Taiwanese: Chinese. 2) Every place may belong to China. 3) They call Guangdongen ancestors. And so on…By the way, I wonder why The Chinese will love to discriminate against everyone. They have nothing to be proud in the world. In turn, everyone has the right to discriminate against the Chinese.

Well, while Myanmar's army dropped the bombs into China. The Chinese were afraid of Myanmar's army invading China. So, don't talk about North Korea, this country makes China wet the Chinese pants. Say, Chinese, do you fear that some ones take China down apart? Only one thing the Chinese can do is paying money to the other countries. See, I don't understand why Taiwan fears the fucking China. Looking at South China Sea, there have naval ships cruising with aircraft carriers. In the southwest of China, India is close to there. Tibet shall be liberated. At the same time, Russia shall be close to Sinkiang. Tell me now, Chinese, do you know that Russia needs the Northwest of China? In Manchuria, Japan will go back at there. Lala, where does China go?

Oh, let me say these words again: the Chinese contribute nothing but shits to the world. Certainly, the Chinese have no right to write anything about me.

Hongkong, which is sad for itself, makes Hongkongers think deeply. When you analyze its result you should discover that Hongkongers lost their chance over 150 years. If they chose English to be their main language for their daily lives they should be different now. Fortunately, it's a good lesson: while you make a choice you must pay it a price. This is a fair life. No one will pity you because it's superfluous. Maybe Hongkongers never care what they chose. Just like them for the fucking economy, their way is the real estate. So, a wrong way they get and go. In my mind, they could rise by the software industry, but they gave up English in their lives.

Okay now, to my people Dianbaien, when you follow me to choose English (or the other languages if you like) to write everything in your lives, that it means: you have a great life anywhere and anytime. Maybe you don't know that the language is one of most important things in our world. But listen to me again, you must forget the fucking Chinese Characters. That's not our language. I wish you young fellows at schools don't fight for the fucking examination. It's useless. You just want to do what you want to be. Don't be afraid of any predicaments, in front of which it makes you stronger than all the times. Indeed, I was not a brilliant student at schools. I didn't listen to a few teachers when they were teaching in the classrooms. So that I didn't care what they talked. The crazy thing was that: at junior school, I wanted to be a teacher. Then after a few years, I wanted to be a news reporter. I hoped I could study at Jinan University, but shit, I had no chance for it. Tell me, my young fellows, what are your dreams? But when one day you look back for something with a sweet smile, congrats, you're successful now.

I love writing and all things of mathematics. To me, that's super crazy. And one thing you must know is that: the writing and the mathematics that both choose me, I have no place to escape.

Sure, I once studied at a fucking college for the fucking medical information of the fucking computer. At there, I was always studying in the fucking library for the fucking reading and for the fucking writing. I was mad for literature. I didn't mind of other things. In last year, I left out of the fucking college to go my way. But my parents still don't know that. Sometimes, I think, my parents will kill me. I'm not a good son. Haha.

Exactly, when I had finished some stories in the fucking Chinese, the tragedy was coming. No place published them but Hongkong. That matter made me so furious that God damns China all. Until one day I sold all Chinese books to the waste collector, I was laughing, and said: fuck off, Chinese; fuck off, all Chinese books.

I love the sea. I also love the pop music and the great movies. You guess what do I say at next time? You just keep calm and pay me money. So, don't tell me that you are a fucking physicist; or you are a fucking mathematician; or you are a fucking writer. Why? That's serious. Of course, you might say someone in your heart is an idol, but you don't know that they were the little roles in the human history.

Ha, over one billion Chinese in the world is one of the biggest jokes. So, If you are Chinese you must have enough courage to live in the world.

The other superfluous words I don't want to talk.

1: $$\frac{1}{x^2}+\frac{1}{y^2}=z^2 \text{ , or } x^{-2}+y^{-2}=z^2 \text{ , or } x^2+y^2=x^2 y^2 z^2 \text{ ,}$$

$$x>0,y>0,z>0 \text{ .}$$

Let $z^2=2p$, it has: $\begin{cases}\dfrac{1}{x^2}+d=p=\dfrac{z^2}{2}\\[2mm]\dfrac{1}{x^2}+2d=\dfrac{1}{y^2}\end{cases}$, d is the common difference.

$\dfrac{1}{x^2}$, $\dfrac{z^2}{2}$ and $\dfrac{1}{y^2}$ are the arithmetic progression.

Then let $z=2^n$, $(n=1,2,\dots,\infty)$, and $d=\dfrac{1}{x^2}$, that it means
$$\begin{cases} \dfrac{1}{x^2}+\dfrac{1}{x^2}=\dfrac{(2^n)^2}{2} \\ \dfrac{1}{x^2}+\dfrac{2}{x^2}=\dfrac{1}{y^2} \end{cases}$$

.

Because $\dfrac{1}{x^2}+\dfrac{1}{x^2}=\dfrac{(2^n)^2}{2}$, in which $4=x^2(2^n)^2$, and $x=\dfrac{2}{2^n}$.

Also because $\dfrac{1}{x^2}+\dfrac{2}{x^2}=\dfrac{1}{y^2}$, in which $3y^2=x^2$, and $y=\dfrac{2\sqrt{3}}{32^n}$.

Now, just let the base number 2 to be any prime number P ,

these have: $x=\dfrac{2}{p^n}$, $y=\dfrac{2\sqrt{3}}{3p^n}$, $z=p^n$.

So when $\dfrac{1}{x^2}+\dfrac{1}{y^2}=z^2$, $x=\dfrac{2}{p^n}$, $y=\dfrac{2\sqrt{3}}{3p^n}$, $z=p^n$, $(n=1,2,3,\dots,\infty)$,

and p is any prime number, I call it X-F Theorem 1.

A series of X-F Theorem 1

1) $\dfrac{1}{ax^2}+\dfrac{1}{y^2}=z^2$, $a>0,x>0,y>0,z>0$.

when $a=(p+1)^2$, these have:

$x=\dfrac{2}{(p+1)p^n}$, $y=\dfrac{2\sqrt{3}}{3p^n}$, $z=p^n$, $(n=1,2,3,\dots,\infty)$,

and p is any prime number.

2) $\dfrac{1}{x^2}+\dfrac{1}{b\,y^2}=z^2$, $b>0, x>0, y>0, >0$.

when $b=(p+2)^2$, these have:

$$x=\dfrac{2}{p^n} \ , \quad y=\dfrac{2\sqrt{3}}{3(p+2)\,p^n} \ , \quad z=p^n \ , \quad (n=1,2,3,\ldots,\infty) \ ,$$

and p is any prime number.

3) $\dfrac{1}{x^2}+\dfrac{1}{y^2}=c\,z^2$, $c>0, x>0, y>0, z>0$.

when $c=(p+3)^2$, these have:

$$x=\dfrac{2}{(P+3)\,p^n} \ , \quad y=\dfrac{2\sqrt{3}}{3(p+3)\,p^n} \ , \quad z=p^n \ , \quad (n=1,2,3,\ldots,\infty) \ ,$$

and p is any prime number.

4) $\dfrac{1}{a\,x^2}+\dfrac{1}{b\,y^2}=z^2$, $a>0, b>0, x>0, y>0, z>0$.

when $a=(p+1)^2$, $b=(p+2)^2$, these have:

$$x=\dfrac{2}{(p+1)\,P^n} \ , \quad y=\dfrac{2\sqrt{3}}{3(p+2)\,p^n} \ , \quad z=p^n \ , \quad (n=1,2,3,\ldots,\infty) \ ,$$

and p is any prime number.

5) $\dfrac{1}{a\,x^2}+\dfrac{1}{y^2}=c\,z^2$, $a>0,c>0,x>0,y>0,z>0$.

when $a=(p+1)^2$, $c=(p+3)^2$, these have:

$$x=\dfrac{2}{(p+1)(p+3)\,p^2}\ ,\quad y=\dfrac{2\sqrt{3}}{3(p+3)\,p^n}\ ,\quad z=p^n\ ,\quad (n=1,2,3,\ldots,\infty)\ ,$$

and p is any prime number.

6) $\dfrac{1}{x^2}+\dfrac{1}{b\,y^2}=c\,z^2$, $b>0,c>0,x>0,y>0,z>0$.

when $b=(p+2)^2$, $c=(p+3)^2$, these have:

$$x=\dfrac{2}{(p+3)\,p^n}\ ,\quad y=\dfrac{2\sqrt{3}}{(p+2)(p+3)\,p^2}\ ,\quad z=p^n\ ,\quad (n=1,2,3,\ldots,\infty)\ ,$$

and p is any prime number.

7) $\dfrac{1}{a\,x^2}+\dfrac{1}{b\,y^2}=c\,z^2$, $a>0,b>0,c>0,x>0,y>0,z>0$.

when $a=(p+1)^2$, $b=(p+2)^2$, $c=(p+3)^2$, these have:

$$x=\dfrac{2}{(p+1)(p+3)\,p^n}\ ,\quad y=\dfrac{2\sqrt{3}}{3(p+2)(p+3)\,p^n}\ ,\quad z=p^n\ ,\quad (n=1,2,3,\ldots,\infty)\ ,$$

and p is any prime number.

2: $\quad x^2+y^2=\dfrac{1}{z^2}$, $x>0,y>0,z>0$.

Let $\dfrac{1}{z^2}=2p$, it has $\begin{cases} x^2+d=\dfrac{1}{2z^2} \\ x^2+2d=y^2 \end{cases}$, d is the common difference.

x^2 , $\dfrac{1}{2z^2}$ and y^2 are the arithmetic progression.

Then let $z=2^n$, $(n=1,2,3,...,\infty)$, and $d=x^2$,

that it means $\begin{cases} x^2+x^2=\dfrac{1}{2(2^n)^2} \\ x^2+2x^2=y^2 \end{cases}$.

Because $x^2+x^2=\dfrac{1}{2(2^n)^2}$, in which $4x^2(2^n)^2=1$, and $x=\dfrac{1}{22^n}$.

Also because $x^2+2x^2=y^2$, in which $3x^2=y^2$, and $y=\dfrac{\sqrt{3}}{22^n}$.

Now, just let the base number 2 to be any prime number p ,

these have: $x=\dfrac{1}{2p^n}$, $y=\dfrac{\sqrt{3}}{2p^n}$ and $z=p^n$.

So when $x^2+y^2=\dfrac{1}{z^2}$, $x=\dfrac{1}{2p^n}$, $y=\dfrac{\sqrt{3}}{2p^n}$, $z=p^n$, $(n=1,2,3,...,\infty)$,

and p is any prime number, I call it X-F Theorem 2.

A series of X-F Theorem 2

1) $ax^2+y^2=\dfrac{1}{z^2}$, $a>0, x>0, y>0, z>0$.

when $a=(p+1)^2$, these have:

$$x=\frac{1}{2(p+1)p^n} \quad , \quad y=\frac{\sqrt{3}}{2p^n} \quad , \quad z=p^n \quad , \quad (n=1,2,3,...,\infty) \quad ,$$

and p is any prime number.

2) $x^2+b\,y^2=\dfrac{1}{z^2}$, $b>0,x>0,y>0,z>0$,

when $b=(p+2)^2$, these have:

$$x=\frac{1}{2\,p^n} \quad , \quad y=\frac{\sqrt{3}}{2(p+2)p^n} \quad , \quad z=p^n \quad , \quad (n=1,2,3,...,\infty) \quad ,$$

and p is any prime number.

3) $x^2+y^2=\dfrac{1}{c\,z^2}$, $c>0,x>0,y>0,z>0$,

when $c=(p+3)^2$, these have:

$$x=\frac{1}{2(p+3)\,p^n} \quad , \quad y=\frac{\sqrt{3}}{2(p+3)\,p^n} \quad , \quad z=p^n \quad , \quad (n=1,2,3,...,\infty) \quad ,$$

and p is any prime number.

4) $a\,x^2+b\,y^2=\dfrac{1}{z^2}$, $a>0,b>0,x>0,y>0,z>0$,

when $a=(p+1)^2$, $b=(p+2)^2$, these have:

$$x=\frac{1}{2(p+1)\,p^n} \quad , \quad y=\frac{\sqrt{3}}{2(p+2)\,p^n} \quad , \quad z=p^n \quad , \quad (n=1,2,3,...,\infty) \quad ,$$

and p is any prime number.

5) $a x^2 + y^2 = \dfrac{1}{c z^2}$, $a>0, c>0, x>0, y>0, z>0$,

when $a=(p+1)^2$, $c=(p+3)^2$, these have:

$$x = \dfrac{1}{2(p+1)(p+3) p^n} \quad , \quad y = \dfrac{\sqrt{3}}{2(p+3) p^n} \quad , \quad z=p^n \ , \quad (n=1,2,3,...,\infty) \ ,$$

and p is any prime number.

6) $x^2 + b y^2 = \dfrac{1}{c z^2}$, $b>0, c>0, x>0, y>0, z>0$,

when $b=(p+2)^2$, $c=(p+3)^2$, these have:

$$x = \dfrac{1}{2(p+3) p^n} \quad , \quad y = \dfrac{\sqrt{3}}{2(p+2)(p+3) p^n} \quad , \quad z=p^n \ , \quad (n=1,2,3,...,\infty) \ ,$$

and p is any prime number.

7) $a x^2 + b y^2 = \dfrac{1}{c z^2}$, $a>0, b>0, c>0, x>0, y>0, z>0$,

when $a=(p+1)^2$, $b=(p+2)^2$, $c=(p+3)^2$, these have:

$$x = \dfrac{1}{2(p+1)(p+3) p^n} \quad , \quad y = \dfrac{\sqrt{3}}{2(p+2)(p+3) p^n} \quad , \quad z=p^n \ , \quad (n=1,2,3,...,\infty) \ ,$$

and p is any prime number.

3:

$$\frac{1}{x^2}+y^2=z^2 \ , \quad x>0, y>0, z>0 \ .$$

Let $z^2=2p$, it has $\begin{cases} \dfrac{1}{x^2}+d=\dfrac{z^2}{2} \\ \dfrac{1}{x^2}+2d=y^2 \end{cases}$, d is the common difference.

$\dfrac{1}{x^2}$, $\dfrac{z^2}{2}$ and y^2 are the arithmetic progression.

Then let $z=2(2^n)$, $(n=1,2,3,...,\infty)$, $d=\dfrac{1}{x^2}$,

that it means $\begin{cases} \dfrac{1}{x^2}+\dfrac{1}{x^2}=\dfrac{2^2(2^n)^2}{2} \\ \dfrac{1}{x^2}+\dfrac{2}{x^2}=y^2 \end{cases}$.

Because $\dfrac{1}{x^2}+\dfrac{1}{x^2}=\dfrac{2^2(2^n)^2}{2}$, in which $4=x^2 2^2(2^n)^2$, and $x=\dfrac{1}{2^n}$.

Also because $\dfrac{1}{x^2}+\dfrac{2}{x^2}=y^2$, in which $3=x^2 y^2$, and $y=\sqrt{3}2^n$.

Now, just let the base number 2 to be any prime number p ,

these have: $x=\dfrac{1}{p^n}$, $y=\sqrt{3}p^n$, and $z=2p^n$.

So when $\dfrac{1}{x^2}+y^2=z^2$, $x=\dfrac{1}{p^n}$, $y=\sqrt{3}p^n$, $z=2p^n$, $(n=1,2,3,...,\infty)$,

and p is any prime number, I call it X-F Theorem 3.

A series of X-F Theorem 3

1) $\dfrac{1}{a\,x^2}+y^2=z^2$, $a>0,x>0,y>0,z>0$,

when $a=(2p+1)^2$, these have:

$$x=\dfrac{1}{(2p+1)p^n} \quad,\quad y=\sqrt{3}\,p^n \quad,\quad z=2\,p^n \quad,\quad (n=1,2,3,...,\infty) \quad,$$

and p is any prime number.

2) $\dfrac{1}{x^2}+b\,y^2=z^2$, $b>0,x>0,y>0,z>0$,

when $b=(2p+2)^2$, these have:

$$x=\dfrac{1}{p^n} \quad,\quad y=\dfrac{\sqrt{3}\,p^n}{2p+2} \quad,\quad z=2\,p^n \quad,\quad (n=1,2,3,...,\infty) \quad,$$

and p is any prime number.

3) $\dfrac{1}{x^2}+y^2=c\,z^2$, $c>0,x>0,y>0,z>0$,

when $c=(2p+3)^2$, these have:

$$x=\dfrac{1}{(2p+3)\,p^n} \quad,\quad y=\sqrt{3}(2p+3)\,p^n \quad,\quad z=2\,p^n \quad,\quad (n=1,2,3,...,\infty) \quad,$$

and p is any prime number.

4) $\dfrac{1}{a\,x^2}+b\,y^2=z^2$, $a>0,b>0,x>0,y>0,z>0$,

when $a=(2\,p+1)^2$, $b=(2\,p+2)^2$, these have:

$x=\dfrac{1}{(2\,p+1)\,p^n}$, $y=\dfrac{\sqrt{3}\,p^n}{2\,p+2}$, $z=2\,p^n$, $(n=1,2,3,...,\infty)$,

and p is any prime number.

5) $\dfrac{1}{a\,x^2}+y^2=c\,z^2$, $a>0,c>0,x>0,y>0,z>0$,

when $a=(2\,p+1)^2$, $c=(2\,p+3)^2$, these have:

$x=\dfrac{1}{(2\,p+1)(2\,p+3)\,p^n}$, $y=\sqrt{3}(2\,p+3)\,p^n$, $z=2\,p^n$, $(n=1,2,3,...,\infty)$,

and p is any prime number.

6) $\dfrac{1}{x^2}+b\,y^2=c\,z^2$, $b>0,c>0,x>0,y>0,z>0$,

when $b=(2\,p+2)^2$, $c=(2\,p+3)^2$, these have:

$x=\dfrac{1}{(2\,p+3)\,p^n}$, $y=\dfrac{\sqrt{3}(2\,p+3)\,p^n}{(2\,p+2)}$, $z=2\,p^n$, $(n=1,2,3,...,\infty)$,

and p is any prime number.

7) $\dfrac{1}{a\,x^2}+b\,y^2=c\,z^2$, $a>0,b>0,c>0,x>0,y>0,z>0$,

when $a=(2\,p+1)^2$, $b=(2\,p+2)^2$, $c=(2\,p+3)^2$, these have:

$$x = \frac{1}{(2p+1)(2p+3)p^n} \quad , \quad y = \frac{\sqrt{3}(2p+3)p^n}{2p+2} \quad , \quad z = 2p^n \quad , \quad (n=1,2,3,...,\infty) \quad ,$$

and p is any prime number.

4:

$$\frac{1}{x^2} + y^2 = \frac{1}{z^2} \quad , \quad x>0, y>0, z>0 \quad .$$

Let $\frac{1}{z^2} = 2p$, it has
$$\begin{cases} \frac{1}{x^2} + d = \frac{1}{2z^2} \\ \frac{1}{x^2} + 2d = y^2 \end{cases} \quad , \quad d \text{ is the common difference.}$$

$\frac{1}{x^2}$, $\frac{1}{2z^2}$ and y^2 are the arithmetic progression.

Then let $z = 2^n$, $(n=1,2,3,...,)$, $d = \frac{1}{x^2}$,

that it means
$$\begin{cases} \frac{1}{x^2} + \frac{1}{x^2} = \frac{1}{2(2^n)^2} \\ \frac{1}{x^2} + \frac{2}{x^2} = y^2 \end{cases} \quad .$$

Because $\frac{1}{x^2} + \frac{1}{x^2} = \frac{1}{2(2^n)^2}$, in which $x^2 = 4(2^n)^2$, and $x = 2(2^n)$.

Also because $\frac{1}{x^2} + \frac{2}{x^2} = y^2$, in which $3 = x^2 y^2$, and $y = \frac{\sqrt{3}}{2(2^n)}$.

Now, just let the base number 2 to be any prime number p ,

these have: $x=2p^n$, $y=\dfrac{\sqrt{3}}{2p^n}$, and $z=p^n$.

So when $\dfrac{1}{x^2}+y^2=\dfrac{1}{z^2}$, $x=2p^n$, $y=\dfrac{\sqrt{3}}{2p^n}$, $z=p^n$, $(n=1,2,3,...,\infty)$,

and p is any prime number, I call it X-F Theorem 4.

A series of X-F Theorem 4

1) $\dfrac{1}{ax^2}+y^2=\dfrac{1}{z^2}$, $a>0,x>0,y>0,z>0$,

when $a=(3p+1)^2$, these have:

$x=\dfrac{2p^n}{3p+1}$, $y=\dfrac{\sqrt{3}}{2p^n}$, $z=p^n$, $(n=1,2,3,...,\infty)$,

and p is any prime number.

2) $\dfrac{1}{x^2}+by^2=\dfrac{1}{z^2}$, $b>0,x>0,y>0,z>0$,

when $b=(3p+2)^2$, these have:

$x=2p^n$, $y=\dfrac{\sqrt{3}}{2(3p+2)p^n}$, $z=p^n$, $(n=1,2,3,...,\infty)$,

and p is any prime number.

3) $\dfrac{1}{x^2}+y^2=\dfrac{1}{cz^2}$, $c>0,x>0,y>0,z>0$,

when $c=(3p+3)^2$, these have:

$$x = 2(3p+3)p^n \quad , \quad y = \frac{\sqrt{3}}{2(3p+3)p^n} \quad , \quad z = p^n \quad , \quad (n=1,2,3,\ldots,\infty) \ ,$$

and p is any prime number.

4) $\quad \dfrac{1}{a\,x^2} + b\,y^2 = \dfrac{1}{z^2} \quad , \quad a>0, b>0, x>0, y>0, z>0 \ ,$

when $\quad a = (3p+1)^2 \quad , \quad b = (3p+2)^2 \ ,$ these have:

$$x = \frac{2\,p^n}{3p+1} \quad , \quad y = \frac{\sqrt{3}}{2(3p+2)p^n} \quad , \quad z = p^n \quad , \quad (n=1,2,3,\ldots,\infty) \ ,$$

and p is any prime number.

5) $\quad \dfrac{1}{a\,x^2} + y^2 = \dfrac{1}{c\,z^2} \quad , \quad a>0, c>0, x>0, y>0, z>0 \ ,$

when $\quad a = (3p+1)^2 \quad , \quad b = (3p+3)^2 \ ,$ these have:

$$x = \frac{2(3p+3)\,p^n}{3p+1} \quad , \quad y = \frac{\sqrt{3}}{2(3p+3)p^n} \quad , \quad z = p^n \quad , \quad (n=1,2,3,\ldots,\infty) \ ,$$

and p is any prime number.

6) $\quad \dfrac{1}{x^2} + b\,y^2 = \dfrac{1}{c\,z^2} \quad , \quad b>0, c>0, x>0, y>0, z>0 \ ,$

when $\ b = (3p+2)^2 \quad , \quad c = (3p+3)^2 \ ,$ these have:

$$x = 2(3p+3)p^n \quad , \quad y = \frac{\sqrt{3}}{2(3p+2)(3p+3)p^n} \quad , \quad z = p^n \quad (n=1,2,3,\ldots,\infty) \ ,$$

and p is any prime number.

7) $\dfrac{1}{a\,x^2}+b\,y^2=\dfrac{1}{c\,z^2}$, $a>0,b>0,c>0,x>0,y>0,z>0$,

when $a=(3p+1)^2$, $b=(3p+2)^2$, $c=(3p+3)^2$, these have:

$x=\dfrac{2(3p+3)\,p^n}{3p+1}$, $y=\dfrac{\sqrt{3}}{2(3p+2)(3p+3)\,p^n}$, $z=p^n$, $(n=1,2,3,\dots,\infty)$,

and p is any prime number.

5:

$$x^{\frac{1}{2}}+y^{\frac{1}{2}}=z^{\frac{1}{2}} \ , \quad x>0,\,y>0,\,z>0 \ .$$

Let $z^{\frac{1}{2}}=2p$, it has $\begin{cases} x^{\frac{1}{2}}+d=\dfrac{z^{\frac{1}{2}}}{2} \\[2mm] x^{\frac{1}{2}}+2d=y^{\frac{1}{2}} \end{cases}$, d is the common difference.

$x^{\frac{1}{2}}$, $\dfrac{z^{\frac{1}{2}}}{2}$ and $y^{\frac{1}{2}}$ are the arithmetic progression.

Then let $z=2$, $d=x^{\frac{1}{2}}$,

that it means: $\begin{cases} x^{\frac{1}{2}}+x^{\frac{1}{2}}=\dfrac{2^{\frac{1}{2}}}{2} \\[2mm] x^{\frac{1}{2}}+2x^{\frac{1}{2}}=y^{\frac{1}{2}} \end{cases}$.

Because $x^{\frac{1}{2}}+x^{\frac{1}{2}}=\dfrac{2^{\frac{1}{2}}}{2}$, in which $4x^{\frac{1}{2}}=2^{\frac{1}{2}}$, and $x=\dfrac{2}{16}$.

Also because $x^{\frac{1}{2}}+2x^{\frac{1}{2}}=y^{\frac{1}{2}}$, in which $3x^{\frac{1}{2}}=y^{\frac{1}{2}}$, and $y=\dfrac{9}{16}2$.

Now, just let the number 2 to be any prime number p ,

these have: $x=\dfrac{p}{16}$, $y=\dfrac{9p}{16}$, and $z=p$.

So when $x^{\frac{1}{2}}+y^{\frac{1}{2}}=z^{\frac{1}{2}}$, $x=\dfrac{p}{16}$, $y=\dfrac{9p}{16}$, $z=p$,

and p is any prime number, I call it X-F Theorem 5.

A series of X-F Theorem 5

1) $ax^{\frac{1}{2}}+y^{\frac{1}{2}}=z^{\frac{1}{2}}$, $a>0, x>0, y>0, z>0$,

   when $a=p+1$, these have:

   $x=\dfrac{p}{16(p+1)^2}$, $y=\dfrac{9p}{16}$, $z=p$,

   and p is any prime number.

2) $x^{\frac{1}{2}}+by^{\frac{1}{2}}=z^{\frac{1}{2}}$, $b>0, x>0, y>0, z>0$,

   when $b=p+2$, these have:

   $x=\dfrac{p}{16}$, $y=\dfrac{9p}{16(p+2)^2}$, $z=p$,

   and p is any prime number.

3) $x^{\frac{1}{2}}+y^{\frac{1}{2}}=c\,z^{\frac{1}{2}}$, $c>0, x>0, y>0, z>0$,

   when $c=p+3$, these have:

$$x=\frac{(p+3)^2\,p}{16} \quad , \quad y=\frac{9(p+3)^2\,p}{16} \quad , \quad z=p \ ,$$

   and p is any prime number.

4) $a\,x^{\frac{1}{2}}+b\,y^{\frac{1}{2}}=z^{\frac{1}{2}}$, $a>0, b>0, x>0, y>0, z>0$,

   when $a=p+1$, $b=p+2$, these have:

$$x=\frac{p}{16(p+1)^2} \quad , \quad y=\frac{9\,p}{16(p+2)^2} \quad , \quad z=p \ ,$$

   and p is any prime number.

5) $a\,x^{\frac{1}{2}}+y^{\frac{1}{2}}=c\,z^{\frac{1}{2}}$, $a>0, c>0, x>0, y>0, z>0$,

   when $a=p+1$, $c=p+3$, these have:

$$x=\frac{(p+3)^2\,p}{16(p+1)^2} \quad , \quad y=\frac{9(p+3)^2\,p}{16} \quad , \quad z=p \ ,$$

   and p is any prime number.

6) $x^{\frac{1}{2}}+b\,y^{\frac{1}{2}}=c\,z^{\frac{1}{2}}$, $b>0, c>0, x>0, y>0, z>0$,

   when $b=p+2$, $c=p+3$, these have:

$$x=\frac{(p+3)^2 p}{16} \quad , \quad y=\frac{9(p+3)^2 p}{16(p+2)^2} \quad , \quad z=p \quad ,$$

and p is any prime number.

7) $\quad a x^{\frac{1}{2}}+b y^{\frac{1}{2}}=c z^{\frac{1}{2}} \quad , \quad a>0,b>0,c>0,x>0,y>0,z>0 \quad ,$

when $\quad a=p+1 \quad , \quad b=p+2 \quad , \quad c=p+3 \quad$, these have:

$$x=\frac{(p+3)^2 p}{16(p+1)^2} \quad , \quad y=\frac{9(p+3)^2 p}{16(p+2)^2} \quad , \quad z=p \quad ,$$

and p is any prime number.

6:

$$x^{\frac{1}{n}}+y^{\frac{1}{n}}=z^{\frac{1}{n}} \quad , \quad x>0,y>0,z>0 \quad , \quad (n=3,4,\dots,\infty) \quad ,$$

Let $\quad z^{\frac{1}{n}}=2p \quad$, it has: $\quad \begin{cases} x^{\frac{1}{n}}+d=\dfrac{z^{\frac{1}{n}}}{2} \\ x^{\frac{1}{n}}+2d=y^{\frac{1}{n}} \end{cases} \quad . \quad d$ is the common difference.

$x^{\frac{1}{n}} \quad , \quad \dfrac{z^{\frac{1}{n}}}{2} \quad$ and $\quad y^{\frac{1}{n}} \quad$ are the arithmetic progression.

Then let $\quad z=2 \quad , \quad d=x^{\frac{1}{n}} \quad ,$

That it means:
$$\begin{cases} x^{\frac{1}{n}}+x^{\frac{1}{n}}=\dfrac{2^{\frac{1}{n}}}{2} \\ x^{\frac{1}{n}}+2x^{\frac{1}{n}}=y^{\frac{1}{n}} \end{cases}.$$

Because $x^{\frac{1}{n}}+x^{\frac{1}{n}}=\dfrac{2^{\frac{1}{n}}}{2}$, in which $4x^{\frac{1}{n}}=2^{\frac{1}{n}}$, and $x=2^{(n)\log_{(2)}(\frac{1}{4})+1}$.

Also because $x^{\frac{1}{n}}+2x^{\frac{1}{n}}=y^{\frac{1}{n}}$, in which $3x^{\frac{1}{n}}=y^{\frac{1}{n}}$, and $y=2^{(n)\log_{(2)}(\frac{3}{4})+1}$.

Now, just let the base number 2 to be any prime number p ,

these have: $x=p^{(n)\log_{(p)}(\frac{1}{4})+1}$, $y=p^{(n)\log_{(p)}(\frac{3}{4})+1}$, $z=p$, $(n=3,4,\dots,\infty)$,

and p is any prime number, I call it X-F Theorem 6.

A series of X-F Theorem 6

1) $ax^{\frac{1}{n}}+y^{\frac{1}{n}}=z^{\frac{1}{n}}$, $a>0,x>0,y>0,z>0$,

when $a=p+1$, these have:

$x=p^{(n)\log_{(p)}(\frac{1}{4p+4})+1}$, $y=p^{(n)\log_{(p)}(\frac{3}{4})+1}$, $z=p$, $(n=3,4,\dots,\infty)$,

and p is any prime number.

2) $ax^{\frac{1}{n}}+by^{\frac{1}{n}}=z^{\frac{1}{n}}$, $a>0,b>0,x>0,y>0,z>0$,

when $a=p+1$, $b=p+2$, these have:

$$x=p^{(n)\log_{(p)}(\frac{1}{4p+4})+1} \quad , \quad y=p^{(n)\log_{(p)}(\frac{3}{4p+8})+1} \quad , \quad z=p \quad , \quad (n=3,4,\ldots,\infty) \quad ,$$

and p is any prime number.

3) $\quad x^{\frac{1}{n}}+b\,y^{\frac{1}{n}}=z^{\frac{1}{n}} \quad , \quad b>0, x>0, y>0, z>0 \quad ,$

when $b=p+2$,these have:

$$x=p^{(n)\log_{(p)}(\frac{1}{4})+1} \quad , \quad y=p^{(n)\log_{(p)}(\frac{3}{4p+8})+1} \quad , \quad z=p \quad , \quad (n=3,4,\ldots,\infty) \quad ,$$

and p is any prime number.

4) $\quad x^{\frac{1}{n}}+y^{\frac{1}{n}}=c\,z^{\frac{1}{n}} \quad , \quad c>0, x>0, y>0, z>0 \quad ,$

when $c=p+3$, these have:

$$x=p^{(n)\log_{(p)}(\frac{p+3}{4})+1} \quad , \quad y=p^{(n)\log_{(p)}(\frac{3p+9}{4})+1} \quad , \quad z=p \quad , \quad (n=3,4,\ldots,\infty) \quad ,$$

and p is any prime number.

5) $\quad a\,x^{\frac{1}{n}}+y^{\frac{1}{n}}=c\,z^{\frac{1}{n}} \quad , \quad a>0, c>0, x>0, y>0, z>0 \quad ,$

when $a=p+1$, $c=p+3$, these have:

$$x=p^{(n)\log_{(p)}(\frac{p+3}{4p+4})+1} \quad , \quad y=p^{(n)\log_{(p)}(\frac{3p+9}{4})+1} \quad , \quad z=p \quad , \quad (n=1,2,3,\ldots,\infty) \quad ,$$

and p is any prime number.

6) $\quad x^{\frac{1}{n}}+b\,y^{\frac{1}{n}}=c\,z^{\frac{1}{n}} \quad , \quad b>0, c>0, x>0, y>0, z>0 \quad ,$

when $b=p+2$, $c=p+3$, these have:

$$x = p^{(n)\log_{(p)}(\frac{p+3}{4})+1} \quad , \quad y = p^{(n)\log_{(p)}(\frac{3p+9}{4p+8})+1} \quad , \quad z = p \quad , \quad (n=3,4,\ldots,\infty) \ ,$$

and p is any prime number.

7) $\quad a\,x^{\frac{1}{n}} + b\,y^{\frac{1}{n}} = c\,z^{\frac{1}{n}} \quad , \quad a>0, b>0, c>0, x>0, y>0, z>0 \ ,$

when $a=p+1$, $b=p+2$, $c=p+3$, these have

$$x = p^{(n)\log_{(p)}(\frac{p+3}{4p+4})+1} \quad , \quad y = p^{(n)\log_{(p)}(\frac{3p+9}{4p+8})+1} \quad , \quad z = p \quad , \quad (n=3,4,\ldots,\infty) \ ,$$

and p is any prime number.

7:

$$x^{\frac{1}{k}} + y^{\frac{1}{m}} = z^{\frac{1}{j}} \quad , \quad k>0, m>0, j>0, x>0, y>0, z>0 \ ,$$

Let $z^{\frac{1}{j}} = 2p$, it has: $\quad \begin{cases} x^{\frac{1}{k}} + d = \dfrac{z^{\frac{1}{j}}}{2} \\[4mm] x^{\frac{1}{k}} + 2d = y^{\frac{1}{m}} \end{cases}$, d is the common difference.

$x^{\frac{1}{k}}$, $\dfrac{z^{\frac{1}{j}}}{2}$ and $y^{\frac{1}{m}}$ are the arithmetic progression.

Then let $z=2$, $d=x^{\frac{1}{k}}$, and $j=n$, $k=n+1$, $m=n+2$,

that it means: $\quad \begin{cases} x^{\frac{1}{n+1}} + x^{\frac{1}{n+1}} = \dfrac{2^{\frac{1}{n}}}{2} \\[4mm] x^{\frac{1}{n+1}} + 2x^{\frac{1}{n+1}} = y^{\frac{1}{n+2}} \end{cases}$,

Because $\ x^{\frac{1}{n+1}}+x^{\frac{1}{n+1}}=\dfrac{2^{\frac{1}{n}}}{2}\ $, in which $\ 4x^{\frac{1}{n+1}}=2^{\frac{1}{n}}\ $, and $\ x=2^{(n+1)\log_{(2)}(\frac{1}{4})+\frac{n+1}{n}}$.

Also because $\ x^{\frac{1}{n+1}}+2x^{\frac{1}{n+1}}=y^{\frac{1}{n+2}}$,

in which $\ 3x^{\frac{1}{n+1}}=y^{\frac{1}{n+2}}\ $, $\ y=2^{(n+2)\log_{(2)}(\frac{3}{4})+\frac{n+2}{n}}$.

Now, just let the base number 2 to be any prime number $\ p$,

these have : $\ x=p^{(n+1)\log_{(p)}(\frac{1}{4})+\frac{n+1}{n}}\ $, $\ y=p^{(n+2)\log_{(p)}(\frac{3}{4})+\frac{n+2}{n}}\ $, $\ z=p$.

So when $\ x^{\frac{1}{k}}+y^{\frac{1}{m}}=z^{\frac{1}{j}}\ $, $\ k>0,m>0,j>0,x>0,y>0,z>0$,

$k=n+1\ $, $\ m=n+2\ $, $\ z=n\ $, $\ (n=1,2,3,...,\infty)$,

$x=p^{(n+1)\log_{(p)}(\frac{1}{4})+\frac{n+1}{n}}\ $, $\ y=p^{(n+2)\log_{(p)}(\frac{3}{4})+\frac{n+2}{n}}\ $, $\ z=p$,

and $\ p$ is any prime number, I call it X-F Theorem 7.

A series of X-F Theorem 7

1) $\ ax^{\frac{1}{k}}+y^{\frac{1}{m}}=z^{\frac{1}{j}}\ $, $\ a>0,k>0,m>0,j>0,x>0,y>0,z>0$,

when $\ a=2p+1\ $, $\ k=n+1\ $, $\ m=n+2\ $, $\ j=n\ $, $\ (n=1,2,3,...,\infty)$,

$x=p^{(n+1)\log_{(p)}(\frac{1}{8p+4})+\frac{n+1}{n}}\ $, $\ y=p^{(n+2)\log_{(p)}(\frac{3}{4})+\frac{n+2}{n}}\ $, $\ z=p$,

and $\ p$ is any prime number.

2) $x^{\frac{1}{k}}+b\,y^{\frac{1}{m}}=z^{\frac{1}{j}}$, $b>0, x>0, y>0, z>0, k>0, m>0, j>0$,

when $b=2p+2, k=n+1, m=n+2, j=n$, $(n=1,2,3,\ldots,\infty)$,

$$x=p^{(n+1)\log_{(p)}(\frac{1}{4})+\frac{n+1}{n}} \quad , \quad y=p^{(n+2)\log_{(p)}(\frac{3}{8p+8})+\frac{n+2}{n}} \quad , \quad z=p \ ,$$

and p is any prime number.

3) $x^{\frac{1}{k}}+y^{\frac{1}{m}}=c\,z^{\frac{1}{j}}$, $c>0, x>0, y>0, z>0, k>0, m>0, j>0$,

when $c=2p+3, k=n+1, m=n+2, j=n$, $(n=1,2,3,\ldots,\infty)$, these have:

$$x=p^{(n+1)\log_{(p)}(\frac{2p+3}{4})+\frac{n+1}{n}} \quad , \quad y=p^{(n+2)\log_{(p)}(\frac{6p+9}{4})+\frac{n+2}{n}} \quad , \quad z=p \ ,$$

and p is any prime number.

4) $a\,x^{\frac{1}{k}}+b\,y^{\frac{1}{m}}=z^{\frac{1}{j}}$, $a>0, b>0, x>0, y>0, z>0, k>0, m>0, j>0$,

When $a=2p+1, b=2p+2, k=n+1, m=n+2, j=n$, $(n=1,2,3,\ldots,\infty)$,

these have: $x=p^{(n+1)\log_{(p)}(\frac{1}{8p+4})+\frac{n+1}{n}} \quad , \quad y=p^{(n+2)\log_{(p)}(\frac{3}{8p+8})+\frac{n+2}{n}}$,

$z=p$, and p is any prime number.

5) $a\,x^{\frac{1}{k}}+y^{\frac{1}{m}}=c\,z^{\frac{1}{j}}$, $a>0, c>0, x>0, y>0, z>0$,

When $a=2p+1, c=2p+3, k=n+1, m=n+2, j=n$, $(n=1,2,3,\ldots,\infty)$,

these have: $x=p^{(n+1)\log_{(p)}(\frac{2p+3}{8p+4})+\frac{n+1}{n}} \quad , \quad y=p^{(n+2)\log_{(p)}(\frac{6p+9}{4})+\frac{n+2}{n}}$,

$z=p$, and p is any prime number.

6) $x^{\frac{1}{k}}+b\,y^{\frac{1}{m}}=c\,z^{\frac{1}{j}}$, $b>0,c>0,x>0,y>0,z>0,k>0,m>0,j>0$,

When $b=2p+2,c=2p+3,k=n+1,m=n+2,j=n$, $(n=1,2,3,...,\infty)$,

these have: $x=p^{(n+1)\log_p(\frac{2p+3}{4})+\frac{n+1}{n}}$, $y=p^{(n+2)\log_{(p)}(\frac{6p+9}{8p+8})+\frac{n+2}{n}}$,

$z=p$, and p is any prime number.

7) $a\,x^{\frac{1}{k}}+b\,y^{\frac{1}{m}}=c\,z^{\frac{1}{j}}$, $a>0,b>0,c>0,x>0,y>0,z>0,k>0,m>0,j>0$,

When $a=2p+1,b=2p+2,c=2p+3,k=n+1,m=n+2,j=n$,

$(n=1,2,3,...,\infty)$, these have:

$x=p^{(n+1)\log_{(p)}(\frac{2p+3}{8p+4})+\frac{n+1}{n}}$, $y=p^{(n+2)\log_{(p)}(\frac{6p+9}{8p+8})+\frac{n+2}{n}}$,

$z=p$, and p is any prime number.

8:

$x^{\frac{e}{k}}+y^{\frac{f}{m}}=z^{\frac{g}{j}}$, $e>0,f>0,g>0,k>0,m>0,j>0,x>0,y>0,z>0$,

Let $z^{\frac{g}{j}}=2p$, it has: $\begin{cases} x^{\frac{e}{k}}+d=\dfrac{z^{\frac{g}{j}}}{2} \\ x^{\frac{e}{k}}+2d=y^{\frac{f}{m}} \end{cases}$, d is the common difference.

$x^{\frac{e}{k}}$, $\dfrac{z^{\frac{g}{j}}}{2}$ and $y^{\frac{f}{m}}$ are the arithmetic progression.

Then let $z=2$, $d=x^{\frac{e}{k}}$,

and $k=n+1, m=n+2, j=n, e=n+3, f=n+4, g=n+5$, $(n=1,2,3,\ldots,\infty)$,

that it means:
$$\begin{cases} x^{\frac{n+3}{n+1}} + x^{\frac{n+3}{n+1}} = \dfrac{2^{\frac{n+5}{n}}}{2} \\ x^{\frac{n+3}{n+1}} + 2x^{\frac{n+3}{n+1}} = y^{\frac{n+4}{n+2}} \end{cases}.$$

Because $x^{\frac{n+3}{n+1}} + x^{\frac{n+3}{n+1}} = \dfrac{2^{\frac{n+5}{n}}}{2}$, in which $4x^{\frac{n+3}{n+1}} = 2^{\frac{n+5}{n}}$,

and $x = 2^{(\frac{n+1}{n+3})\log_{(2)}(\frac{1}{4}) + \frac{(n+1)(n+5)}{n(n+3)}}$.

Also because $x^{\frac{n+3}{n+1}} + 2x^{\frac{n+3}{n+1}} = y^{\frac{n+4}{n+2}}$, in which $3x^{\frac{n+3}{n+1}} = y^{\frac{n+4}{n+2}}$,

and $y = 2^{(\frac{n+2}{n+4})\log_{(2)}(\frac{3}{4}) + \frac{(n+2)(n+5)}{n(n+4)}}$.

now, just let the base number 2 to be any prime number p,

these have: $x = p^{(\frac{n+1}{n+3})\log_{(p)}(\frac{1}{4}) + \frac{(n+1)(n+5)}{n(n+3)}}$, $y = p^{(\frac{n+2}{n+4})\log_{(p)}(\frac{3}{4}) + \frac{(n+2)(n+5)}{n(n+4)}}$,

$z = p$.

So when $x^{\frac{e}{k}} + y^{\frac{f}{m}} = z^{\frac{g}{j}}$, $e>0, f>0, g>0, k>0, m>0, j>0, x>0, y>0, z>0$,

$k=n+1, m=n+2, j=n, e=n+3, f=n+4, g=n+5$, $(n=1,2,3,\ldots,\infty)$,

$x = p^{(\frac{n+1}{n+3})\log_{(p)}(\frac{1}{4}) + \frac{(n+1)(n+5)}{n(n+3)}}$, $y = p^{(\frac{n+2}{n+4})\log_{(p)}(\frac{3}{4}) + \frac{(n+2)(n+5)}{n(n+4)}}$, $z=p$,

and p is any prime number, I call it X-F Theorem 8.

A series of X-F Theorem 8

1) $a x^{\frac{e}{k}}+y^{\frac{f}{m}}=z^{\frac{g}{j}}$, $a>0,e>0,f>0,g>0,k>0,m>0,j>0,x>0,y>0,z>0$,

When $a=p+1,k=n+1,m=n+2,j=n,e=n+3,f=n+4,g=n+5$,

$$\left(n=1,2,3,\ldots,\infty\right) ,$$

These have: $x=p^{(\frac{n+1}{n+3})\log_{(p)}(\frac{1}{4p+4})+\frac{(n+1)(n+5)}{n(n+3)}}$, $y=p^{(\frac{n+2}{n+4})\log_{(p)}(\frac{3}{4})+\frac{(n+2)(n+5)}{n(n+4)}}$,

$z=p$, and p is any prime number.

2) $x^{\frac{e}{k}}+b y^{\frac{f}{m}}=z^{\frac{g}{j}}$, $b>0,e>0,f>0,g>0,k>0,m>0,j>0,x>0,y>0,z>0$,

When $b=p+2,k=n+1,m=n+2,j=n,e=n+3,f=n+4,g=n+5$,

$$\left(n=1,2,3,\ldots,\infty\right) ,$$

These have: $x=p^{(\frac{n+1}{n+3})\log_{(p)}(\frac{1}{4})+\frac{(n+1)(n+5)}{n(n+3)}}$, $y=p^{(\frac{n+2}{n+4})\log_{(p)}(\frac{3}{4p+8})+\frac{(n+2)(n+5)}{n(n+4)}}$,

$z=p$, and p is any prime number.

3) $x^{\frac{e}{k}}+y^{\frac{f}{m}}=c z^{\frac{g}{j}}$, $c>0,e>0,f>0,g>0,k>0,m>0,j>0,x>0,y>0,z>0$,

When $c=p+3,k=n+1,m=n+2,j=n,e=n+3,f=n+4,g=n+5$,

$$\left(n=1,2,3,\ldots,\infty\right) ,$$

These have: $x=p^{(\frac{n+1}{n+3})\log_{(p)}(\frac{p+3}{4})+\frac{(n+1)(n+5)}{n(n+3)}}$,

$$y=p^{(\frac{n+2}{n+4})\log_{(p)}(\frac{3p+9}{4})+\frac{(n+2)(n+5)}{n(n+4)}}$$,

$z=p$, and p is any prime number.

4) $\quad a x^{\frac{e}{k}}+b y^{\frac{f}{m}}=z^{\frac{g}{j}}$, $a>0,b>0,e>0,f>0,g>0,k>0,m>0,j>0,x>0,y>0,z>0$,

When $\quad a=p+1,b=p+2,k=n+1,m=n+2,j=n,e=n+3,f=n+4,g=n+5$,

$$(n=1,2,3,\dots,\infty) \ ,$$

These have: $\quad x=p^{(\frac{n+1}{n+3})\log_{(p)}(\frac{1}{4p+4})+\frac{(n+1)(n+5)}{n(n+3)}}$,

$$y=p^{(\frac{n+2}{n+4})\log_{(p)}(\frac{3}{4p+8})+\frac{(n+2)(n+5)}{n(n+4)}}$$,

$z=p$, and p is any prime number.

5) $\quad a x^{\frac{e}{k}}+y^{\frac{f}{m}}=c z^{\frac{g}{j}}$, $a>0,c>0,e>0,f>0,g>0,k>0,m>0,j>0,x>0,y>0,z>0$,

When $\quad a=p+1,c=p+3,k=n+1,m=n+2,j=n,e=n+3,f=n+4,g=n+5$,

$$(n=1,2,3,\dots,\infty) \ ,$$

These have: $\quad x=p^{(\frac{n+1}{n+3})\log_{(p)}(\frac{p+3}{4p+4})+\frac{(n+1)(n+5)}{n(n+3)}}$,

$$y=p^{(\frac{n+2}{n+4})\log_{(p)}(\frac{3p+9}{4})+\frac{(n+2)(n+5)}{n(n+4)}}$$,

$z=p$, and p is any prime number.

6) $\quad x^{\frac{e}{k}}+b y^{\frac{f}{m}}=c z^{\frac{g}{j}}$, $b>0,c>0,e>0,f>0,g>0,k>0,m>0,j>0,x>0,y>0,z>0$,

When $\quad b=p+2,c=p+3,k=n+1,m=n+2,j=n,e=n+3,f=n+4,g=n+5$,

$$\left(n=1,2,3,\ldots,\infty\right) \ ,$$

These have: $\quad x=p^{\left(\frac{n+1}{n+3}\right)\log_{(p)}\left(\frac{p+3}{4}\right)+\frac{(n+1)(n+5)}{n(n+3)}} \ ,$

$$y=p^{\left(\frac{n+2}{n+4}\right)\log_{(p)}\left(\frac{3p+9}{4p+8}\right)+\frac{(n+2)(n+5)}{n(n+4)}} \ ,$$

$z=p$, and $\quad p$ is any prime number.

7) $\quad a\,x^{\frac{e}{k}}+b\,y^{\frac{f}{m}}=c\,z^{\frac{g}{j}} \ , \quad a>0,b>0,c>0,e>0,f>0,g>0,k>0,m>0,j>0,x>0,y>0,z>0 \ ,$

When $\quad a=p+1,b=p+2,c=p+3,k=n+1,m=n+2,j=n,e=n+3,f=n+4,g=n+5 \ ,$

$$\left(n=1,2,3,\ldots,\infty\right) \ ,$$

These have: $\quad x=p^{\left(\frac{n+1}{n+3}\right)\log_{(p)}\left(\frac{p+3}{4p+4}\right)+\frac{(n+1)(n+5)}{n(n+3)}} \ ,$

$$y=p^{\left(\frac{n+2}{n+4}\right)\log_{(p)}\left(\frac{3p+9}{4p+8}\right)+\frac{(n+2)(n+5)}{n(n+4)}} \ ,$$

$z=p$, and $\quad p$ is any prime number.

9:

$$x^{\frac{1}{m}}+y^{\frac{1}{m}}=z^{\frac{1}{n}} \ , \quad m>0,n>0,x>0,y>0,z>0 \ ,$$

Let $\quad z^{\frac{1}{n}}=2p$, it has: $\quad \begin{cases} x^{\frac{1}{m}}+d=\dfrac{z^{\frac{1}{n}}}{2} \ , \\[2mm] x^{\frac{1}{m}}+2d=y^{\frac{1}{m}} \end{cases}$ $\quad d$ is the common difference.

$x^{\frac{1}{m}}$, $\dfrac{z^{\frac{1}{n}}}{2}$ and $y^{\frac{1}{m}}$ are the arithmetic progression.

Then let $z=2$, $d=x^{\frac{1}{m}}$, and $n=m+5$,

That it means: $\begin{cases} x^{\frac{1}{m}}+x^{\frac{1}{m}}=\dfrac{2^{\frac{1}{m+5}}}{2} \\ x^{\frac{1}{m}}+2x^{\frac{1}{m}}=y^{\frac{1}{m}} \end{cases}$.

Because $x^{\frac{1}{m}}+x^{\frac{1}{m}}=2^{\frac{1}{m+5}}$, in which $4x^{\frac{1}{m}}=2^{\frac{1}{m+5}}$, and $x=2^{(m)\log_{(2)}(\frac{1}{4})+\frac{m}{m+5}}$.

Also because $x^{\frac{1}{m}}+2x^{\frac{1}{m}}=y^{\frac{1}{m}}$, in which $3x^{\frac{1}{m}}=y^{\frac{1}{m}}$, and $y=2^{(m)\log_{(2)}(\frac{3}{4})+\frac{m}{m+5}}$.

Now, just let the base number 2 to be any prime number p ,

These have: $x=p^{(m)\log_{(p)}(\frac{1}{4})+\frac{m}{m+5}}$, $y=p^{(m)\log_{(p)}(\frac{3}{4})+\frac{m}{m+5}}$, $z=p$.

So when $x^{\frac{1}{m}}+y^{\frac{1}{m}}=z^{\frac{1}{n}}$, $m>0,n>0,x>0,y>0,z>0$, $n=m+5$,

$$(m=1,2,3,\ldots,\infty) ,$$

$x=p^{(m)\log_{(p)}(\frac{1}{4})+\frac{m}{m+5}}$, $y=p^{(m)\log_{(p)}(\frac{3}{4})+\frac{m}{m+5}}$, $z=p$,

and p is any prime number, I call it X-F Theorem 9.

A series of X-F Theorem 9

1) $ax^{\frac{1}{m}}+y^{\frac{1}{m}}=z^{\frac{1}{n}}$, $a>0,m>0,n>0,x>0,y>0,z>0$,

When $a=5p,n=m+5$, $(m=1,2,3,\ldots,\infty)$,

These have: $x = p^{(m)\log_{(p)}(\frac{1}{20p}) + \frac{m}{m+5}}$, $y = p^{(m)\log_{(p)}(\frac{3}{4}) + \frac{m}{m+5}}$,

$z = p$, and p is any prime number.

2) $x^{\frac{1}{m}} + b\, y^{\frac{1}{m}} = z^{\frac{1}{n}}$, $b>0, m>0, n>0, x>0, y>0, z>0$,

When $b = 4p, n = m+5$, $(m = 1,2,3,\dots,\infty)$,

These have: $x = p^{(m)\log_{(p)}(\frac{1}{4}) + \frac{m}{m+5}}$, $y = p^{(m)\log_{(p)}(\frac{3}{16p}) + \frac{m}{m+5}}$,

$z = p$, and p is any prime number.

3) $x^{\frac{1}{m}} + y^{\frac{1}{m}} = c\, z^{\frac{1}{n}}$, $c>0, m>0, n>0, x>0, y>0, z>0$,

When $c = 7p, n = m+5$, $(m = 1,2,3,\dots,\infty)$,

These have: $x = p^{(m)\log_{(p)}(\frac{7p}{4}) + \frac{m}{m+5}}$, $y = p^{(m)\log_{(p)}(\frac{21p}{4}) + \frac{m}{m+5}}$,

$z = p$, and p is any prime number.

4) $a\, x^{\frac{1}{m}} + b\, y^{\frac{1}{m}} = z^{\frac{1}{n}}$, $a>0, b>0, m>0, n>0, x>0, y>0, z>0$,

When $a = 5p, b = 4p, n = m+5$, $(m = 1,2,3,\dots,\infty)$,

These have: $x = p^{(m)\log_{(p)}(\frac{1}{20p}) + \frac{m}{m+5}}$, $y = p^{(m)\log_{(p)}(\frac{3}{16p}) + \frac{m}{m+5}}$,

$z = p$, and p is any prime number.

5) $a\, x^{\frac{1}{m}} + y^{\frac{1}{m}} = c\, z^{\frac{1}{n}}$, $a>0, c>0, m>0, n>0, x>0, y>0, z>0$,

When $a = 5p, c = 7p, n = m+5$, $(m = 1,2,3,\dots,\infty)$,

These have: $x = p^{(m)\log_{(p)}(\frac{7p}{20p}) + \frac{m}{m+5}}$, $y = p^{(m)\log_{(p)}(\frac{21p}{4}) + \frac{m}{m+5}}$,

$z = p$, and p is any prime number.

6) $x^{\frac{1}{m}} + b\, y^{\frac{1}{m}} = c\, z^{\frac{1}{n}}$, $b>0, c>0, m>0, n>0, x>0, y>0, z>0$,

When $b = 4p, c = 7p, n = m+5$, $(m = 1,2,3,\ldots,\infty)$,

These have: $x = p^{(m)\log_{(p)}(\frac{7p}{4}) + \frac{m}{m+5}}$, $y = p^{(m)\log_{(p)}(\frac{21p}{16p}) + \frac{m}{m+5}}$,

$z = p$, and p is any prime number.

7) $a\, x^{\frac{1}{m}} + b\, y^{\frac{1}{m}} = c\, z^{\frac{1}{n}}$, $a>0, b>0, c>0, m>0, n>0, x>0, y>0, z>0$,

When $a = 5p, b = 4p, c = 7p, n = m+5$, $(m = 1,2,3,\ldots,\infty)$,

These have: $x = p^{(m)\log_{(p)}(\frac{7p}{20p}) + \frac{m}{m+5}}$, $y = p^{(m)\log_{(p)}(\frac{21p}{16p}) + \frac{m}{m+5}}$,

$z = p$, and p is any prime number.

10:

$x^{\frac{k}{m}} + y^{\frac{j}{m}} = z^{\frac{e}{n}}$, $k>0, j>0, e>0, n>0, m>0, x>0, y>0, z>0$,

Let $z^{\frac{e}{n}} = 2p$, it has: $\begin{cases} x^{\frac{k}{m}} + d = \dfrac{z^{\frac{e}{n}}}{2} \\[2mm] x^{\frac{k}{m}} + 2d = y^{\frac{j}{m}} \end{cases}$, d is the common difference.

$x^{\frac{k}{m}}$, $\dfrac{z^{\frac{e}{n}}}{2}$ and $y^{\frac{j}{m}}$ are the arithmetic progression.

Then let $z=2$, $d=x^{\frac{k}{m}}$,

$m=2n+1, k=3n+1, j=4n+1, e=5n+1, (n=1,2,3,\ldots,\infty)$,

That it means:
$$
\begin{cases}
x^{\frac{3n+1}{2n+1}}+x^{\frac{3n+1}{2n+1}}=\dfrac{2^{\frac{5n+1}{n}}}{2}\\[2mm]
x^{\frac{3n+1}{2n+1}}+2x^{\frac{3n+1}{2n+1}}=y^{\frac{4n+1}{2n+1}}
\end{cases}
$$

Because $x^{\frac{3n+1}{2n+1}}+x^{\frac{3n+1}{2n+1}}=\dfrac{2^{\frac{5n+1}{n}}}{2}$, in which $4x^{\frac{3n+1}{2n+1}}=2^{\frac{5n+1}{n}}$,

$$x=2^{(\frac{2n+1}{3n+1})\log_{(2)}(\frac{1}{4})+\frac{(2n+1)(5n+1)}{(3n+1)n}}$$,

Also because $x^{\frac{3n+1}{2n+1}}+2x^{\frac{3n+1}{2n+1}}=y^{\frac{4n+1}{2n+1}}$, in which $3x^{\frac{3n+1}{2n+1}}=y^{\frac{4n+1}{2n+1}}$,

$$y=2^{(\frac{2n+1}{4n+1})\log_{(2)}(\frac{3}{4})+\frac{(2n+1)(5n+1)}{(4n+1)n}}$$.

Now, just let the base number 2 to be any prime number p ,

These have: $x=p^{(\frac{2n+1}{3n+1})\log_{(p)}(\frac{1}{4})+\frac{(2n+1)(5n+1)}{(3n+1)n}}$, $y=p^{(\frac{2n+1}{4n+1})\log_{(p)}(\frac{3}{4})+\frac{(2n+1)(5n+1)}{(4n+1)n}}$,

$z=p$, and p is any prime number.

So when $x^{\frac{k}{m}}+y^{\frac{j}{m}}=z^{\frac{e}{n}}$, $k>0, j>0, e>0, n>0, m>0, x>0, y>0, z>0$,

$m=2n+1, k=3n+1, j=4n+1, e=5n+1, (n=1,2,3,\ldots,\infty)$,

$$x = p^{(\frac{2n+1}{3n+1})\log_{(p)}(\frac{1}{4})+\frac{(2n+1)(5n+1)}{(3n+1)n}} \quad , \quad y = p^{(\frac{2n+1}{4n+1})\log_{(p)}(\frac{3}{4})+\frac{(2n+1)(5n+1)}{(4n+1)n}} \quad ,$$

$z = p$, and p is any prime number, I call it X-F Theorem 10.

A series of X-F Theorem 10

1) $\quad a\,x^{\frac{k}{m}} + y^{\frac{j}{m}} = z^{\frac{e}{n}}$, $\quad a>0, k>0, j>0, e>0, n>0, m>0, x>0, y>0, z>0$,

When $\quad a=3\,p, m=2n+1, k=3n+1, j=4n+1, e=5n+1, (n=1,2,3,\ldots,\infty)$,

These have: $\quad x = p^{(\frac{2n+1}{3n+1})\log_{(p)}(\frac{1}{12\,p})+\frac{(2n+1)(5n+1)}{(3n+1)n}}$,

$$y = p^{(\frac{2n+1}{4n+1})\log_{(p)}(\frac{3}{4})+\frac{(2n+1)(5n+1)}{(4n+1)n}} \quad ,$$

$\quad z = p$, and p is any prime number.

2) $\quad x^{\frac{k}{m}} + b\,y^{\frac{j}{m}} = z^{\frac{e}{n}}$, $\quad b>0, k>0, j>0, e>0, n>0, m>0, x>0, y>0, z>0$,

When $\quad b=5\,p, m=2n+1, k=3n+1, j=4n+1, e=5n+1, (n=1,2,3,\ldots,\infty)$,

These have: $\quad x = p^{(\frac{2n+1}{3n+1})\log_{(p)}(\frac{1}{4})+\frac{(2n+1)(5n+1)}{(3n+1)n}}$,

$$y = p^{(\frac{2n+1}{4n+1})\log_{(p)}(\frac{3}{20\,p})+\frac{(2n+1)(5n+1)}{(4n+1)n}} \quad ,$$

$\quad z = p$, and p is any prime number.

3) $\quad x^{\frac{k}{m}} + y^{\frac{j}{m}} = c\,z^{\frac{e}{n}}$, $\quad c>0, k>0, j>0, e>0, n>0, m>0, x>0, y>0, z>0$,

When $\quad c=7\,p, m=2n+1, k=3n+1, j=4n+1, e=5n+1, (n=1,2,3,\ldots,\infty)$,

These have: $x = p^{(\frac{2n+1}{3n+1})\log_{(p)}(\frac{7p}{4}) + \frac{(2n+1)(5n+1)}{(3n+1)n}}$,

$$y = p^{(\frac{2n+1}{4n+1})\log_{(p)}(\frac{21p}{4}) + \frac{(2n+1)(5n+1)}{(4n+1)n}}$$,

$z = p$, and p is any prime number.

4) $a x^{\frac{k}{m}} + b y^{\frac{j}{m}} = z^{\frac{e}{n}}$, $a>0, b>0, k>0, j>0, e>0, n>0, m>0, x>0, y>0, z>0$,

When $a=3p, b=5p, m=2n+1, k=3n+1, j=4n+1, e=5n+1, (n=1,2,3,\ldots,\infty)$,

These have: $x = p^{(\frac{2n+1}{3n+1})\log_{(p)}(\frac{1}{12p}) + \frac{(2n+1)(5n+1)}{(3n+1)n}}$,

$$y = p^{(\frac{2n+1}{4n+1})\log_{(p)}(\frac{3}{20p}) + \frac{(2n+1)(5n+1)}{(4n+1)n}}$$,

$z = p$, and p is any prime number.

5) $a x^{\frac{k}{m}} + y^{\frac{j}{m}} = c z^{\frac{e}{n}}$, $a>0, c>0, k>0, j>0, e>0, n>0, m>0, x>0, y>0, z>0$,

When $a=3p, c=7p, m=2n+1, k=3n+1, j=4n+1, e=5n+1, (n=1,2,3,\ldots,\infty)$,

These have: $x = p^{(\frac{2n+1}{3n+1})\log_{(p)}(\frac{7}{12}) + \frac{(2n+1)(5n+1)}{(3n+1)n}}$,

$$y = p^{(\frac{2n+1}{4n+1})\log_{(p)}(\frac{21p}{4}) + \frac{(2n+1)(5n+1)}{(4n+1)n}}$$,

$z = p$, and p is any prime number.

6) $x^{\frac{k}{m}} + b y^{\frac{j}{m}} = c z^{\frac{e}{n}}$, $b>0, c>0, k>0, j>0, e>0, n>0, m>0, x>0, y>0, z>0$,

When $b=5p, c=7p, m=2n+1, k=3n+1, j=4n+1, e=5n+1, (n=1,2,3,\ldots,\infty)$,

These have: $x = p^{(\frac{2n+1}{3n+1})\log_{(p)}(\frac{7p}{4}) + \frac{(2n+1)(5n+1)}{(3n+1)n}}$,

$$y = p^{(\frac{2n+1}{4n+1})\log_{(p)}(\frac{21}{20}) + \frac{(2n+1)(5n+1)}{(4n+1)n}}$$,

$z = p$, and p is any prime number.

7) $a x^{\frac{k}{m}} + b y^{\frac{j}{m}} = c z^{\frac{e}{n}}$, $b>0, b>0, c>0, k>0, j>0, e>0, n>0, m>0, x>0, y>0, z>0$,

When $a = 3p, b = 5p, c = 7p, m = 2n+1, k = 3n+1, j = 4n+1, e = 5n+1, (n=1,2,3,...,\infty)$,

These have: $x = p^{(\frac{2n+1}{3n+1})\log_{(p)}(\frac{7}{12}) + \frac{(2n+1)(5n+1)}{(3n+1)n}}$,

$$y = p^{(\frac{2n+1}{4n+1})\log_{(p)}(\frac{21}{20}) + \frac{(2n+1)(5n+1)}{(4n+1)n}}$$,

$z = p$, and p is any prime number.

11:

$$x^m + y^m = z^n , \quad m>0, n>0, x>0, y>0, z>0 ,$$

Let $z^n = 2p$, it has: $\begin{cases} x^m + d = \dfrac{z^n}{2} \\ x^m + 2d = y^m \end{cases}$, d is the common difference.

x^m , $\dfrac{z^n}{2}$ and y^m are the arithmetic progression.

Then let $z = 2, d = x^m, n = 5m+2, (m=1,2,3,...,\infty)$,

That it means: $\begin{cases} x^m + x^m = \dfrac{2^{(5m+2)}}{2} \\ x^m + 2x^m = y^m \end{cases}$.

Because $x^m + x^m = \dfrac{2^{(5m+2)}}{2}$, in which $4\,x^m = 2^{(5m+2)}$,

and $x = 2^{(\frac{1}{m})\log_{(2)}(\frac{1}{4}) + \frac{5m+2}{m}}$.

Also because $x^m + 2\,x^m = y^m$, in which $3\,x^m = y^m$,

and $y = 2^{(\frac{1}{m})\log_{(2)}(\frac{3}{4}) + \frac{5m+2}{m}}$.

Now, just let the base number 2 to be any prime number p ,

These have: $x = p^{(\frac{1}{m})\log_{(p)}(\frac{1}{4}) + \frac{5m+2}{m}}$, $y = p^{(\frac{1}{m})\log_{(p)}(\frac{3}{4}) + \frac{5m+2}{m}}$,

$z = p$, and p is any prime number.

So when $x^m + y^m = z^n$, $m > 0, n > 0, x > 0, y > 0, z > 0$,

$$n = 5m + 2, (m = 1,2,3,\ldots,\infty) \ ,$$

$$x = p^{(\frac{1}{m})\log_{(p)}(\frac{1}{4}) + \frac{5m+2}{m}} \ , \quad y = p^{(\frac{1}{m})\log_{(p)}(\frac{3}{4}) + \frac{5m+2}{m}} \ , \quad z = p \ ,$$

and p is any prime number, I call it X-F Theorem 11.

A series of X-F Theorem 11

1) $a\,x^m + y^m = z^n$, $a > 0, m > 0, n > 0, x > 0, y > 0, z > 0$,

When $a = p + 5, n = 5m + 2, (m = 1,2,3,\ldots,\infty)$,

These have: $x = p^{(\frac{1}{m})\log_{(p)}(\frac{1}{4p+20}) + \frac{5m+2}{m}}$, $y = p^{(\frac{1}{m})\log_{(p)}(\frac{3}{4}) + \frac{5m+2}{m}}$,

$z=p$, and p is any prime number.

2) $x^m + b\,y^m = z^n$, $b>0, m>0, n>0, x>0, y>0, z>0$,

When $b=p+6, n=5m+2, (m=1,2,3,...,\infty)$,

These have: $x=p^{(\frac{1}{m})\log_{(p)}(\frac{1}{4})+\frac{5m+2}{m}}$, $y=p^{(\frac{1}{m})\log_{(p)}(\frac{3}{4p+24})+\frac{5m+2}{m}}$,

$z=p$, and p is any prime number.

3) $x^m + y^m = c\,z^n$, $c>0, m>0, n>0, x>0, y>0, z>0$,

When $c=p+7, n=5m+2, (m=1,2,3,...,\infty)$,

These have: $x=p^{(\frac{1}{m})\log_{(p)}(\frac{p+7}{4})+\frac{5m+2}{m}}$, $y=p^{(\frac{1}{m})\log_{(p)}(\frac{3p+21}{4})+\frac{5m+2}{m}}$,

$z=p$, and p is any prime number.

4) $a\,x^m + b\,y^m = z^n$, $a>0, b>0, m>0, n>0, x>0, y>0, z>0$,

When $a=p+5, b=p+6, n=5m+2, (m=1,2,3,...,\infty)$,

These have: $x=p^{(\frac{1}{m})\log_{(p)}(\frac{1}{4p+20})+\frac{5m+2}{m}}$, $y=p^{(\frac{1}{m})\log_{(p)}(\frac{3}{4p+24})+\frac{5m+2}{m}}$,

$z=p$, and p is any prime number.

5) $a\,x^m + y^m = c\,z^n$, $a>0, c>0, m>0, n>0, x>0, y>0, z>0$,

When $a=p+5, c=p+7, n=5m+2, (m=1,2,3,...,\infty)$,

These have: $x=p^{(\frac{1}{m})\log_{(p)}(\frac{p+7}{4p+20})+\frac{5m+2}{m}}$, $y=p^{(\frac{1}{m})\log_{(p)}(\frac{3p+21}{4})+\frac{5m+2}{m}}$,

$z=p$, and p is any prime number.

6) $x^m + b\,y^m = c\,z^n$, $b>0, c>0, m>0, n>0, x>0, y>0, z>0$,

When $b=p+6, c=p+7, n=5m+2, (m=1,2,3,...,\infty)$,

These have: $x = p^{(\frac{1}{m})\log_{(p)}(\frac{p+7}{4}) + \frac{5m+2}{m}}$, $y = p^{(\frac{1}{m})\log_{(p)}(\frac{3p+21}{4p+24}) + \frac{5m+2}{m}}$,

$z = p$, and p is any prime number.

7) $a\,x^m + b\,y^m = c\,z^n$, $a>0, b>0, c>0, m>0, n>0, x>0, y>0, z>0$,

When $a=p+5, b=p+6, c=p+7, n=5m+2, (m=1,2,3,...,\infty)$,

These have: $x = p^{(\frac{1}{m})\log_{(p)}(\frac{p+7}{4p+20}) + \frac{5m+2}{m}}$, $y = p^{(\frac{1}{m})\log_{(p)}(\frac{3p+21}{4p+24}) + \frac{5m+2}{m}}$,

$z = p$, and p is any prime number.

12:

$x^m + y^n = z^m$, $m>0, n>0, x>0, y>0, z>0$,

Let $z^m = 2p$, it has: $\begin{cases} x^m + d = \dfrac{z^m}{2} \\ x^m + 2d = y^n \end{cases}$, d is the common difference.

x^m , $\dfrac{z^m}{2}$ and y^n are the arithmetic progression.

Then let $z=2, d=x^m, n=3m+4, (m=1,2,3,...,\infty)$,

That it means: $\begin{cases} x^m + x^m = \dfrac{2^m}{2} \\ x^m + 2x^m = y^{(3m+4)} \end{cases}$.

Because $x^m + x^m = \dfrac{2^m}{2}$, in which $4x^m = 2^m$, and $x = 2^{(\frac{1}{m})\log_{(2)}(\frac{1}{4}) + 1}$.

Also because $x^m + 2x^m = y^{(3m+4)}$, in which $3x^m = y^{(3m+4)}$,

and $y = 2^{(\frac{1}{3m+4})\log_{(2)}(\frac{3}{4}) + \frac{m}{3m+4}}$.

Now, just let the base number 2 to be any prime number p ,

These have: $x = p^{(\frac{1}{m})\log_{(p)}(\frac{1}{4}) + 1}$, $y = p^{(\frac{1}{3m+4})\log_{(p)}(\frac{3}{4}) + \frac{m}{3m+4}}$,

$z = p$, and p is any prime number.

So when $x^m + y^n = z^m$, $m>0, n>0, x>0, y>0, z>0$,

$n = 3m+4, (m=1,2,3,\ldots,\infty)$,

$x = p^{(\frac{1}{m})\log_{(p)}(\frac{1}{4}) + 1}$, $y = p^{(\frac{1}{3m+4})\log_{(p)}(\frac{3}{4}) + \frac{m}{3m+4}}$, $z = p$,

and p is any prime number, I call it X-F Theorem 12.

A series of X-F Theorem 12

1) $ax^m + y^n = z^m$, $a>0, m>0, n>0, x>0, y>0, z>0$,

When $a = 5p+1, n = 3m+4, (m=1,2,3,\ldots,\infty)$,

These have: $x = p^{(\frac{1}{m})\log_{(p)}(\frac{1}{20p+4}) + 1}$, $y = p^{(\frac{1}{3m+4})\log_{(p)}(\frac{3}{4}) + \frac{m}{3m+4}}$,

$z = p$, and p is any prime number.

2) $x^m + by^n = z^m$, $b>0, m>0, n>0, x>0, y>0, z>0$,

When $b = 5p+2, n = 3m+4, (m=1,2,3,\ldots,\infty)$,

These have: $x = p^{(\frac{1}{m})\log_{(p)}(\frac{1}{4})+1}$, $y = p^{(\frac{1}{3m+4})\log_{(p)}(\frac{3}{20p+8})+\frac{m}{3m+4}}$,

$z = p$, and p is any prime number.

3) $x^m + y^n = c\,z^m$, $c > 0, m > 0, n > 0, x > 0, y > 0, z > 0$,

When $c = 5p+3, n = 3m+4, (m = 1,2,3,\dots,\infty)$,

These have: $x = p^{(\frac{1}{m})\log_{(p)}(\frac{5p+3}{4})+1}$, $y = p^{(\frac{1}{3m+4})\log_{(p)}(\frac{15p+9}{4})+\frac{m}{3m+4}}$,

$z = p$, and p is any prime number.

4) $a\,x^m + b\,y^n = z^m$, $a > 0, b > 0, m > 0, n > 0, x > 0, y > 0, z > 0$,

When $a = 5p+1, b = 5p+2, n = 3m+4, (m = 1,2,3,\dots,\infty)$,

These have: $x = p^{(\frac{1}{m})\log_{(p)}(\frac{1}{20p+4})+1}$, $y = p^{(\frac{1}{3m+4})\log_{(p)}(\frac{3}{20p+8})+\frac{m}{3m+4}}$,

$z = p$, and p is any prime number.

5) $a\,x^m + y^n = c\,z^m$, $a > 0, c > 0, m > 0, n > 0, x > 0, y > 0, z > 0$,

When $a = 5p+1, c = 5p+3, n = 3m+4, (m = 1,2,3,\dots,\infty)$,

These have: $x = p^{(\frac{1}{m})\log_{(p)}(\frac{5p+3}{20p+4})+1}$, $y = p^{(\frac{1}{3m+4})\log_{(p)}(\frac{15p+9}{4})+\frac{m}{3m+4}}$,

$z = p$, and p is any prime number.

6) $x^m + b\,y^n = c\,z^m$, $b > 0, c > 0, m > 0, n > 0, x > 0, y > 0, z > 0$,

When $b = 5p+2, c = 5p+3, n = 3m+4, (m = 1,2,3,\dots,\infty)$,

These have: $x = p^{(\frac{1}{m})\log_{(p)}(\frac{5p+3}{4})+1}$, $y = p^{(\frac{1}{3m+4})\log_{(p)}(\frac{15p+9}{20p+8})+\frac{m}{3m+4}}$,

$z = p$, and p is any prime number.

7) $a x^m + b y^n = c z^m$, $a > 0, b > 0, c > 0, m > 0, n > 0, x > 0, y > 0, z > 0$,

When $a = 5p+1, b = 5p+2, c = 5p+3, n = 3m+4, (m = 1,2,3,\ldots,\infty)$,

These have: $x = p^{(\frac{1}{m})\log_{(p)}(\frac{5p+3}{20p+4})+1}$, $y = p^{(\frac{1}{3m+4})\log_{(p)}(\frac{15p+9}{20p+8})+\frac{m}{3m+4}}$,

$z = p$, and p is any prime number.

13:

$$x^{\frac{m}{k}} + y^{\frac{j}{n}} = z^{\frac{m}{n}} \quad , \quad k > 0, m > 0, j > 0, n > 0, x > 0, y > 0, z > 0 ,$$

Let $z^{\frac{m}{n}} = 2p$, it has:
$$\begin{cases} x^{\frac{m}{k}} + d = \dfrac{z^{\frac{m}{n}}}{2} \\ x^{\frac{m}{k}} + 2x^{\frac{m}{k}} = y^{\frac{j}{n}} \end{cases} \quad , \quad d \text{ is the common difference.}$$

$x^{\frac{m}{k}}$, $\dfrac{z^{\frac{m}{n}}}{2}$ and $y^{\frac{j}{n}}$ are the arithmetic progression.

Then let $z = 2, d = x^{\frac{m}{n}}, k = m+3, j = 2m+2, n = 4m+5, (m = 1,2,3,\ldots,\infty)$,

That it means:
$$\begin{cases} x^{\frac{m}{m+3}} + x^{\frac{m}{m+3}} = \dfrac{2^{\frac{m}{4m+5}}}{2} \\ x^{\frac{m}{m+3}} + 2x^{\frac{m}{m+3}} = y^{\frac{2m+2}{4m+5}} \end{cases} .$$

Because $x^{\frac{m}{m+3}} + x^{\frac{m}{m+3}} = \dfrac{2^{\frac{m}{4m+5}}}{2}$, in which $4x^{\frac{m}{m+3}} = 2^{\frac{m}{4m+5}}$, and $\quad x=2^{(\frac{m+3}{m})\log_{(2)}(\frac{1}{4})+\frac{m+3}{4m+5}}$.

Also because $\quad x^{\frac{m}{m+3}}+2x^{\frac{m}{m+3}}=y^{\frac{2m+2}{4m+5}}$, in which $\quad 3x^{\frac{m}{m+3}}=y^{\frac{2m+2}{4m+5}}$,

and $\quad y=2^{(\frac{4m+5}{2m+2})\log_{(2)}(\frac{3}{4})+\frac{m}{2m+2}}$.

Now, just let the base number 2 to be any prime number $\quad p$,

These have: $\quad x=p^{(\frac{m+3}{m})\log_{(p)}(\frac{1}{4})+\frac{m+3}{4m+5}}$, $\quad y=p^{(\frac{4m+5}{2m+2})\log_{(p)}(\frac{3}{4})+\frac{m}{2m+2}}$,

$z=p$, and $\quad p$ is any prime number.

So when $\quad x^{\frac{m}{k}}+y^{\frac{j}{n}}=z^{\frac{m}{n}}$, $\quad k>0, m>0, j>0, n>0, x>0, y>0, z>0$,

$k=m+3, j=2m+2, n=4m+5, (m=1,2,3,...,\infty)$,

$x=p^{(\frac{m+3}{m})\log_{(p)}(\frac{1}{4})+\frac{m+3}{4m+5}}$, $\quad y=p^{(\frac{4m+5}{2m+2})\log_{(p)}(\frac{3}{4})+\frac{m}{2m+2}}$, $\quad z=p$,

and $\quad p$ is any prime number, I call it X-F Theorem 13.

A series of X-F Theorem 13

1) $\quad ax^{\frac{m}{k}}+y^{\frac{j}{n}}=z^{\frac{m}{n}}$, $\quad a>0, k>0, m>0, j>0, n>0, x>0, y>0, z>0$,

When $\quad a=2p+3, k=m+3, j=2m+2, n=4m+5, (m=1,2,3,...,\infty)$,

These have: $\quad x=p^{(\frac{m+3}{m})\log_{(p)}(\frac{1}{8p+12})+\frac{m+3}{4m+5}}$, $\quad y=p^{(\frac{4m+5}{2m+2})\log_{(p)}(\frac{3}{4})+\frac{m}{2m+2}}$,

$z=p$, and $\quad p$ is any prime number.

2) $x^{\frac{m}{k}} + b\, y^{\frac{j}{n}} = z^{\frac{m}{n}}$, $b>0, k>0, m>0, j>0, n>0, x>0, y>0, z>0$,

when $b = 3p+3, k = m+3, j = 2m+2, n = 4m+5, (m=1,2,3,...,\infty)$,

These have: $x = p^{(\frac{m+3}{m})\log_{(p)}(\frac{1}{4}) + \frac{m+3}{4m+5}}$, $y = p^{(\frac{4m+5}{2m+2})\log_{(p)}(\frac{3}{12p+12}) + \frac{m}{2m+2}}$,

$z = p$, and p is any prime number.

3) $x^{\frac{m}{k}} + y^{\frac{j}{n}} = c\, z^{\frac{m}{n}}$, $c>0, k>0, m>0, j>0, n>0, x>0, y>0, z>0$,

When $c = 4p+3, k = m+3, j = 2m+2, n = 4m+5, (m=1,2,3,...,\infty)$,

These have: $x = p^{(\frac{m+3}{m})\log_{(p)}(\frac{4p+3}{4}) + \frac{m+3}{4m+5}}$, $y = p^{(\frac{4m+5}{2m+2})\log_{(p)}(\frac{12p+9}{4}) + \frac{m}{2m+2}}$,

$z = p$, and p is any prime number.

4) $a\, x^{\frac{m}{k}} + b\, y^{\frac{j}{n}} = z^{\frac{m}{n}}$, $a>0, b>0, k>0, m>0, j>0, n>0, x>0, y>0, z>0$,

When $a = 2p+3, b = 3p+3, k = m+3, j = 2m+2, n = 4m+5, (m=1,2,3,...,\infty)$,

These have: $x = p^{(\frac{m+3}{m})\log_{(p)}(\frac{1}{8p+12}) + \frac{m+3}{4m+5}}$, $y = p^{(\frac{4m+5}{2m+2})\log_{(p)}(\frac{3}{12p+12}) + \frac{m}{2m+2}}$,

$z = p$, and p is any prime number.

5) $a\, x^{\frac{m}{k}} + y^{\frac{j}{n}} = c\, z^{\frac{m}{n}}$, $a>0, c>0, k>0, m>0, j>0, n>0, x>0, y>0, z>0$,

When $a = 2p+3, c = 4p+3, k = m+3, j = 2m+2, n = 4m+5, (m=1,2,3,...,\infty)$,

These have: $x = p^{\left(\frac{m+3}{m}\right)\log_{(p)}\left(\frac{4p+3}{8p+12}\right)+\frac{m+3}{4m+5}}$, $y = p^{\left(\frac{4m+5}{2m+2}\right)\log_{(p)}\left(\frac{12p+9}{4}\right)+\frac{m}{2m+2}}$,

$z = p$, and p is any prime number.

6) $x^{\frac{m}{k}} + b\, y^{\frac{j}{n}} = c\, z^{\frac{m}{n}}$, $b>0, c>0, k>0, m>0, j>0, n>0, x>0, y>0, z>0$,

When $b = 3p+3, c = 4p+3, k = m+3, j = 2m+2, n = 4m+5, (m = 1,2,3,\dots,\infty)$,

These have: $x = p^{\left(\frac{m+3}{m}\right)\log_{(p)}\left(\frac{4p+3}{4}\right)+\frac{m+3}{4m+5}}$, $y = p^{\left(\frac{4m+5}{2m+2}\right)\log_{(p)}\left(\frac{12p+9}{12p+12}\right)+\frac{m}{2m+2}}$,

$z = p$, and p is any prime number.

7) $a\, x^{\frac{m}{k}} + b\, y^{\frac{j}{n}} = c\, z^{\frac{m}{n}}$, $a>0, b>0, c>0, k>0, m>0, j>0, n>0, x>0, y>0, z>0$,

When $a = 2p+3, b = 3p+3, c = 4p+3, k = m+3, j = 2m+2, n = 4m+5, (m = 1,2,3,\dots,\infty)$,

These have: $x = p^{\left(\frac{m+3}{m}\right)\log_{(p)}\left(\frac{4p+3}{8p+12}\right)+\frac{m+3}{4m+5}}$, $y = p^{\left(\frac{4m+5}{2m+2}\right)\log_{(p)}\left(\frac{12p+9}{12p+12}\right)+\frac{m}{2m+2}}$,

$z = p$, and p is any prime number.

14:

$x^2 + y^2 = z^{\frac{1}{2}}$, $x>0, y>0, z>0$,

Let $z^{\frac{1}{2}} = 2p$, it has: $\begin{cases} x^2 + d = \dfrac{z^{\frac{1}{2}}}{2} \\ x^2 + 2d = y^2 \end{cases}$, d is the common difference.

x^2 , $\dfrac{z^{\frac{1}{2}}}{2}$ and y^2 are the arithmetic progression.

Then let $z=2, d=x^2$,

That it means: $\begin{cases} x^2+x^2=\dfrac{2^{\frac{1}{2}}}{2} \\ x^2+2x^2=y^2 \end{cases}$.

Because $x^2+x^2=\dfrac{2^{\frac{1}{2}}}{2}$, in which $4x^2=2^{\frac{1}{2}}$, and $x=(\dfrac{1}{2})2^{\frac{1}{4}}$.

Also because $x^2+2x^2=y^2$, in which $3x^2=y^2$, and $y=(\dfrac{\sqrt{3}}{2})2^{\frac{1}{4}}$.

Now, just let the base number 2 to be any prime number p ,

These have: $x=(\dfrac{1}{2})p^{\frac{1}{4}}$, $y=(\dfrac{\sqrt{3}}{2})p^{\frac{1}{4}}$, $z=p$.

So when $x^2+y^2=z^{\frac{1}{2}}$, $x>0, y>0, z>0$,

$$x=(\dfrac{1}{2})p^{\frac{1}{4}} , \quad y=(\dfrac{\sqrt{3}}{2})p^{\frac{1}{4}} , \quad z=p ,$$

and p is any prime number, I call it X-F Theorem 14.

A series of X-F Theorem 14

1) $ax^2+y^2=z^{\frac{1}{2}}$, $a>0, x>0, y>0, z>0$,

When $a=(7p+2)^2$,

These have: $x=(\dfrac{1}{14\,p+4})\,p^{\frac{1}{4}}$, $y=(\dfrac{\sqrt{3}}{2})\,p^{\frac{1}{4}}$, $z=p$,

and p is any prime number.

2) $x^2+b\,y^2=z^{\frac{1}{2}}$, $b>0,x>0,y>0,z>0$,

When $b=(7\,p+3)^2$,

These have: $x=(\dfrac{1}{2})\,p^{\frac{1}{4}}$, $y=(\dfrac{\sqrt{3}}{14\,p+6})\,p^{\frac{1}{4}}$, $z=p$,

and p is any prime number.

3) $x^2+y^2=c\,z^{\frac{1}{2}}$, $c>0,x>0,y>0,z>0$,

When $c=(7\,p+4)^2$,

These have: $x=(\dfrac{7\,p+4}{2})\,p^{\frac{1}{4}}$, $y=\dfrac{\sqrt{3}\,(7\,p+4)}{2}\,p^{\frac{1}{4}}$, $z=p$,

and p is any prime number.

4) $a\,x^2+b\,y^2=z^{\frac{1}{2}}$, $a>0,b>0,x>0,y>0,z>0$,

When $a=(7\,p+2)^2,b=(7\,p+3)^2$,

These have: $x=(\dfrac{1}{14\,p+4})\,p^{\frac{1}{4}}$, $y=(\dfrac{\sqrt{3}}{14\,p+6})\,p^{\frac{1}{4}}$, $z=p$,

and p is any prime number.

5) $\quad a x^2 + y^2 = c z^{\frac{1}{2}}$, $\quad a>0, c>0, x>0, y>0, z>0$,

When $\quad a=(7p+2)^2, c=(7p+4)^2$,

These have: $\quad x=\left(\dfrac{7p+4}{14p+4}\right)p^{\frac{1}{4}}$, $\quad y=\dfrac{\sqrt{3}(7p+4)}{2}p^{\frac{1}{4}}$, $\quad z=p$,

and p is any prime number.

6) $\quad x^2 + b y^2 = c z^{\frac{1}{2}}$, $\quad b>0, c>0, x>0, y>0, z>0$,

When $\quad b=(7p+3)^2, c=(7p+4)^2$,

These have: $\quad x=\left(\dfrac{7p+4}{2}\right)p^{\frac{1}{4}}$, $\quad y=\dfrac{\sqrt{3}(7p+4)}{14p+6}p^{\frac{1}{4}}$, $\quad z=p$,

and p is any prime number.

7) $\quad a x^2 + b y^2 = c z^{\frac{1}{2}}$, $\quad a>0, b>0, c>0, x>0, y>0, z>0$,

When $\quad a=(7p+2)^2, b=(7p+3)^2, c=(7p+4)^2$,

These have: $\quad x=\left(\dfrac{7p+4}{14p+4}\right)p^{\frac{1}{4}}$, $\quad y=\dfrac{\sqrt{3}(7p+4)}{14p+6}p^{\frac{1}{4}}$, $\quad z=p$,

and p is any prime number.

15:

$$x^{\frac{1}{2}} + y^{\frac{1}{2}} = z^2 \ , \quad x>0, y>0, z>0 \ ,$$

Let $z^2 = 2p$, it has:
$$\begin{cases} x^{\frac{1}{2}} + d = \dfrac{z^2}{2} \\ x^{\frac{1}{2}} + 2d = y^{\frac{1}{2}} \end{cases} \ , \quad d \text{ is the common difference.}$$

$x^{\frac{1}{2}}$, $\dfrac{z^2}{2}$ and $y^{\frac{1}{2}}$ are the arithmetic progression.

Then let $z=2, d = x^{\frac{1}{2}}$,

That it means:
$$\begin{cases} x^{\frac{1}{2}} + x^{\frac{1}{2}} = \dfrac{2^2}{2} \\ x^{\frac{1}{2}} + 2x^{\frac{1}{2}} = y^{\frac{1}{2}} \end{cases} \ .$$

Because $x^{\frac{1}{2}} + x^{\frac{1}{2}} = \dfrac{2^2}{2}$, in which $4x^{\frac{1}{2}} = 2^2$, and $x = (\dfrac{1}{16})2^4$.

Also because $x^{\frac{1}{2}} + 2x^{\frac{1}{2}} = y^{\frac{1}{2}}$, in which $3x^{\frac{1}{2}} = y^{\frac{1}{2}}$, and $y = (\dfrac{9}{16})2^4$.

Now, just let the base number 2 to be any prime number p ,

These have: $x = (\dfrac{1}{16})p^4$, $y = (\dfrac{9}{16})p^4$, $z = p$.

So when $x^{\frac{1}{2}} + y^{\frac{1}{2}} = z^2$, $x>0, y>0, z>0$,

$$x = (\dfrac{1}{16})p^4 \ , \quad y = (\dfrac{9}{16})p^4 \ , \quad z = p \ ,$$

and p is any prime number, I call it X-F Theorem 15.

A series of X-F Theorem 15

1) $a x^{\frac{1}{2}} + y^{\frac{1}{2}} = z^2$, $a>0, x>0, y>0, z>0$,

When $a = p+5$,

These have: $x = (\dfrac{1}{4p+20})^2 p^4$, $y = (\dfrac{9}{16}) p^4$, $z = p$,

and p is any prime number.

2) $x^{\frac{1}{2}} + b y^{\frac{1}{2}} = z^2$, $b>0, x>0, y>0, z>0$,

When $b = p+6$,

These have: $x = (\dfrac{1}{16})^2 p^4$, $y = (\dfrac{9}{4p+24}) p^4$, $z = p$,

and p is any prime number.

3) $x^{\frac{1}{2}} + y^{\frac{1}{2}} = c z^2$, $c>0, x>0, y>0, z>0$,

When $c = p+7$,

These have: $x = \dfrac{(p+7)^2}{16} p^4$, $y = \dfrac{9(p+7)^2}{16} p^4$, $z = p$,

and p is any prime number.

4) $\quad a x^{\frac{1}{2}} + b y^{\frac{1}{2}} = z^2$, $\quad a>0, b>0, x>0, y>0, z>0$,

When $\quad a=p+5, b=p+6$,

These have: $\quad x=\dfrac{1}{(4p+20)^2} p^4$, $\quad y=\dfrac{9}{(4p+24)^2} p^4$, $\quad z=p$,

and $\quad p$ is any prime number.

5) $\quad a x^{\frac{1}{2}} + y^{\frac{1}{2}} = c z^2$, $\quad a>0, c>0, x>0, y>0, z>0$,

When $\quad a=p+5, c=p+7$,

These have: $\quad x=\dfrac{(p+7)^2}{(4p+20)^2} p^4$, $\quad y=\dfrac{9(p+7)^2}{16} p^4$, $\quad z=p$,

and $\quad p$ is any prime number.

6) $\quad x^{\frac{1}{2}} + b y^{\frac{1}{2}} = c z^2$, $\quad b>0, c>0, x>0, y>0, z>0$,

When $\quad b=p+6, c=p+7$,

These have: $\quad x=\dfrac{(p+7)^2}{16} p^4$, $\quad y=\dfrac{9(p+7)^2}{(4p+24)^2} p^4$, $\quad z=p$,

and $\quad p$ is any prime number.

7) $\quad a x^{\frac{1}{2}} + b y^{\frac{1}{2}} = c z^2$, $\quad a>0, b>0, c>0, x>0, y>0, z>0$,

When $a = p+5, b = p+6, c = p+7$,

These have: $x = \dfrac{(p+7)^2}{(4p+20)^2} p^4$, $y = \dfrac{9(p+7)^2}{(4p+24)^2} p^4$, $z = p$,

and p is any prime number.

16:

$$x^2 + y^{\frac{1}{2}} = z^2 , \quad x>0, y>0, z>0 ,$$

Let $z^2 = 2p$, it has:
$$\begin{cases} x^2 + d = \dfrac{z^2}{2} \\ x^2 + 2d = y^{\frac{1}{2}} \end{cases}$$
, d is the common difference.

x^2 , $\dfrac{z^2}{2}$ and $y^{\frac{1}{2}}$ are the arithmetic progression.

Then let $z = 2, d = x^2$,

That it means:
$$\begin{cases} x^2 + x^2 = \dfrac{2^2}{2} \\ x^2 + 2x^2 = y^{\frac{1}{2}} \end{cases}.$$

Because $x^2 + x^2 = \dfrac{2^2}{2}$, in which $4x^2 = 2^2$, and $x = (\dfrac{1}{2})2$.

Also because $x^2 + 2x^2 = y^{\frac{1}{2}}$, in which $3x^2 = y^{\frac{1}{2}}$, and $y = (\dfrac{9}{16})2^4$.

Now, just let the base number 2 to be any prime number p ,

These have: $x=(\frac{1}{2})p$, $y=(\frac{9}{16})p^4$, $z=p$.

So when $x^2+y^{\frac{1}{2}}=z^2$, $x>0, y>0, z>0$,

$$x=(\frac{1}{2})p \; , \; y=(\frac{9}{16})p^4 \; , \; z=p \; ,$$

and p is any prime number, I call it X-F Theorem 16.

A series of X-F Theorem 16

1) $ax^2+y^{\frac{1}{2}}=z^2$, $a>0, x>0, y>0, z>0$,

When $a=p^2$,

These have: $x=(\frac{1}{2p})p \; , \; y=(\frac{9}{16})p^4 \; , \; z=p \; ,$

and p is any prime number.

2) $x^2+by^{\frac{1}{2}}=z^2$, $b>0, x>0, y>0, z>0$,

When $b=p$,

These have: $x=(\frac{1}{2})p \; , \; y=(\frac{3}{4p})^2 p^4 \; , \; z=p \; ,$

and p is any prime number.

3) $x^2+y^{\frac{1}{2}}=cz^2$, $c>0, x>0, y>0, z>0$,

When $c=(p+1)^2$,

These have: $x=(\dfrac{p+1}{2})p$, $y=\dfrac{9(p+1)^4}{16}p^4$, $z=p$,

and p is any prime number.

4) $ax^2+by^{\frac{1}{2}}=z^2$, $a>0,b>0,x>0,y>0,z>0$,

When $a=p^2,b=p$,

These have: $x=(\dfrac{1}{2p})p$, $y=(\dfrac{9}{16p^2})p^4$, $z=p$,

and p is any prime number.

5) $ax^2+y^{\frac{1}{2}}=cz^2$, $a>0,c>0,x>0,y>0,z>0$,

When $a=p^2,c=(p+1)^2$,

These have: $x=(\dfrac{p+1}{2p})p$, $y=\dfrac{9(p+1)^4}{16}p^4$, $z=p$,

and p is any prime number.

6) $x^2+by^{\frac{1}{2}}=cz^2$, $b>0,c>0,x>0,y>0,z>0$,

When $b=p,c=(p+1)^2$,

These have: $x=(\dfrac{p+1}{2})p$, $y=\dfrac{9(p+1)^4}{16\,p^2}p^4$, $z=p$,

and p is any prime number.

7) $a\,x^2+b\,y^{\frac{1}{2}}=c\,z^2$, $a>0,b>0,c>0,x>0,y>0,z>0$,

When $a=p^2,b=p,c=(p+1)^2$,

These have: $x=(\dfrac{p+1}{2\,p})p$, $y=\dfrac{9(p+1)^4}{16\,p^2}p^4$, $z=p$,

and p is any prime number.

17:

$$x^2+y^{\frac{1}{2}}=z^{\frac{1}{2}} , \quad x>0,y>0,z>0 ,$$

Let $z^{\frac{1}{2}}=2\,p$, it has: $\begin{cases} x^2+d=\dfrac{z^{\frac{1}{2}}}{2} , \\ x^2+2\,d=y^{\frac{1}{2}} \end{cases}$, d is the common difference.

x^2 , $\dfrac{z^{\frac{1}{2}}}{2}$, and $y^{\frac{1}{2}}$ are the arithmetic progression.

Then let $z=2, d=x^2$,

That it means:
$$\begin{cases} x^2+x^2=\dfrac{2^{\frac{1}{2}}}{2} \\[2mm] x^2+2x^2=y^{\frac{1}{2}} \end{cases}.$$

Because $x^2+x^2=\dfrac{2^{\frac{1}{2}}}{2}$, in which $4x^2=2^{\frac{1}{2}}$, and $x=(\dfrac{1}{2})2^{\frac{1}{4}}$.

Also because $x^2+2x^2=y^{\frac{1}{2}}$, in which $3x^2=y^{\frac{1}{2}}$, and $y=(\dfrac{9}{16})2$.

Now, just let the base number 2 to be any prime number p ,

These have: $x=(\dfrac{1}{2})p^{\frac{1}{4}}$, $y=(\dfrac{9}{16})p$, $z=p$.

So when $x^2+y^{\frac{1}{2}}=z^{\frac{1}{2}}$, $x>0, y>0, z>0$,

$x=(\dfrac{1}{2})p^{\frac{1}{4}}$, $y=(\dfrac{9}{16})p$, $z=p$,

and p is any prime number, I call it X-F Theorem 17.

A series of X-F Theorem 17

1) $ax^2+y^{\frac{1}{2}}=z^{\frac{1}{2}}$, $a>0, x>0, y>0, z>0$,

When $a=p^2$,

These have: $x=(\dfrac{1}{2p})p^{\frac{1}{4}}$, $y=(\dfrac{9}{16})p$, $z=p$,

and p is any prime number.

2) $x^2 + b\, y^{\frac{1}{2}} = z^{\frac{1}{2}}$, $b>0, x>0, y>0, z>0$,

When $b = p+1$,

These have: $x = \left(\dfrac{1}{2}\right) p^{\frac{1}{4}}$, $y = \dfrac{9}{16(p+1)^2} p$, $z = p$,

and p is any prime number.

3) $x^2 + y^{\frac{1}{2}} = c\, z^{\frac{1}{2}}$, $c>0, x>0, y>0, z>0$,

When $c = (p+2)^2$,

These have: $x = \dfrac{p+2}{2} p^{\frac{1}{4}}$, $y = \dfrac{9(p+2)^4}{16} p$, $z = p$,

and p is any prime number.

4) $a\, x^2 + b\, y^{\frac{1}{2}} = z^{\frac{1}{2}}$, $a>0, b>0, x>0, y>0, z>0$,

When $a = p^2, b = p+1$,

These have: $x = \left(\dfrac{1}{2p}\right) p^{\frac{1}{4}}$, $y = \dfrac{9}{16(p+1)^2} p$, $z = p$,

and p is any prime number.

5) $a x^2 + y^{\frac{1}{2}} = c z^{\frac{1}{2}}$, $a>0, c>0, x>0, y>0, z>0$,

When $a = p^2, c = (p+2)^2$,

These have: $x = \left(\frac{p+2}{2p}\right) p^{\frac{1}{4}}$, $y = \frac{9(p+2)^4}{16} p$, $z = p$,

and p is any prime number.

6) $x^2 + b y^{\frac{1}{2}} = c z^{\frac{1}{2}}$, $b>0, c>0, x>0, y>0, z>0$,

When $b = p+1, c = (p+2)^2$,

These have: $x = \frac{p+2}{2} p^{\frac{1}{4}}$, $y = \frac{9(p+2)^4}{16(p+1)^2} p$, $z = p$,

and p is any prime number.

7) $a x^2 + b y^{\frac{1}{2}} = c z^{\frac{1}{2}}$, $a>0, b>0, c>0, x>0, y>0, z>0$,

When $a = p^2, b = p+1, c = (p+2)^2$,

These have: $x = \left(\frac{p+2}{2p}\right) p^{\frac{1}{4}}$, $y = \frac{9(p+2)^4}{16(p+1)^2} p$, $z = p$,

and p is any prime number.

18:

$$x^{\frac{1}{n}} + y^{\frac{1}{n}} = z^n \ , \quad n>0, x>0, y>0, z>0 \ ,$$

Let $z^n = 2p$, it has: $\begin{cases} x^{\frac{1}{n}} + d = \dfrac{z^n}{2} \\ x^{\frac{1}{n}} + 2d = y^{\frac{1}{n}} \end{cases}$, d is the common difference.

$x^{\frac{1}{n}}$, $\dfrac{z^n}{2}$ and $y^{\frac{1}{n}}$ are the arithmetic progression.

Then let $z=2, d=x^{\frac{1}{n}}$.

That it means: $\begin{cases} x^{\frac{1}{n}} + x^{\frac{1}{n}} = \dfrac{2^n}{2} \\ x^{\frac{1}{n}} + 2x^{\frac{1}{n}} = y^{\frac{1}{n}} \end{cases}$.

Because $x^{\frac{1}{n}} + x^{\frac{1}{n}} = \dfrac{2^n}{2}$, in which $4x^{\frac{1}{n}} = 2^n$, and $x = 2^{(n)\log_{(2)}(\frac{1}{4}) + n^2}$.

Also because $x^{\frac{1}{n}} + 2x^{\frac{1}{n}} = y^{\frac{1}{n}}$, in which $3x^{\frac{1}{n}} = y^{\frac{1}{n}}$, and $y = 2^{(n)\log_{(2)}(\frac{3}{4}) + n^2}$.

Now, just let the base number 2 to be any prime number p ,

These have: $x = p^{(n)\log_{(p)}(\frac{1}{4}) + n^2}$, $y = p^{(n)\log_{(p)}(\frac{3}{4}) + n^2}$, $z = p$.

So when $x^{\frac{1}{n}} + y^{\frac{1}{n}} = z^n$, $n>0, x>0, y>0, z>0$,

$$x = p^{(n)\log_{(p)}(\frac{1}{4}) + n^2} \ , \quad y = p^{(n)\log_{(p)}(\frac{3}{4}) + n^2} \ , \quad z = p \ ,$$

and p is any prime number, I call it X-F Theorem 18.

A series of X-F Theorem 18

1) $a x^{\frac{1}{n}} + y^{\frac{1}{n}} = z^n$, $a>0, n>0, x>0, y>0, z>0$,

   When $a = 5p$,

   These have: $x = p^{(n)\log_{(p)}(\frac{1}{20p})+n^2}$, $y = p^{(n)\log_{(p)}(\frac{3}{4})+n^2}$, $z = p$,

   and p is any prime number.

2) $x^{\frac{1}{n}} + b y^{\frac{1}{n}} = z^n$, $b>0, n>0, x>0, y>0, z>0$,

   When $b = 4p$,

   These have: $x = p^{(n)\log_{(p)}(\frac{1}{4})+n^2}$, $y = p^{(n)\log_{(p)}(\frac{3}{16p})+n^2}$, $z = p$,

   and p is any prime number.

3) $x^{\frac{1}{n}} + y^{\frac{1}{n}} = c z^n$, $c>0, n>0, x>0, y>0, z>0$,

   When $c = p$,

   These have: $x = p^{(n)\log_{(p)}(\frac{p}{4})+n^2}$, $y = p^{(n)\log_{(p)}(\frac{3p}{4})+n^2}$, $z = p$,

   and p is any prime number.

4) $\quad a\,x^{\frac{1}{n}}+b\,y^{\frac{1}{n}}=z^{n}$, $\quad a>0,b>0,n>0,x>0,y>0,z>0$,

When $\quad a=5\,p, b=4\,p$,

These have: $\quad x=p^{(n)\log_{(p)}(\frac{1}{20\,p})+n^2}$, $\quad y=p^{(n)\log_{(p)}(\frac{3}{16\,p})+n^2}$, $\quad z=p$,

and $\quad p$ is any prime number.

5) $\quad a\,x^{\frac{1}{n}}+y^{\frac{1}{n}}=c\,z^{n}$, $\quad a>0,c>0,n>0,x>0,y>0,z>0$,

When $\quad a=5\,p, c=p$,

These have: $\quad x=p^{(n)\log_{(p)}(\frac{1}{20})+n^2}$, $\quad y=p^{(n)\log_{(p)}(\frac{3\,p}{4})+n^2}$, $\quad z=p$,

and $\quad p$ is any prime number.

6) $\quad x^{\frac{1}{n}}+b\,y^{\frac{1}{n}}=c\,z^{n}$, $\quad b>0,c>0,n>0,x>0,y>0,z>0$,

When $\quad b=4\,p, c=p$,

These have: $\quad x=p^{(n)\log_{(p)}(\frac{p}{4})+n^2}$, $\quad y=p^{(n)\log_{(p)}(\frac{3}{16})+n^2}$, $\quad z=p$,

and $\quad p$ is any prime number.

7) $\quad a\,x^{\frac{1}{n}}+b\,y^{\frac{1}{n}}=c\,z^{n}$, $\quad a>0,b>0,c>0,n>0,x>0,y>0,z>0$,

When $a=5p, b=4p, c=p$,

These have: $x=p^{(n)\log_{(p)}(\frac{1}{20})+n^2}$, $y=p^{(n)\log_{(p)}(\frac{3}{16})+n^2}$, $z=p$,

and p is any prime number.

<h2>19:</h2>

$$x^{\frac{k}{n}}+y^{\frac{m}{n}}=z^n \ , \quad k>0, m>0, n>0, x>0, y>0, z>0 \ ,$$

Let $z^n=2p$, it has:
$$\begin{cases} x^{\frac{k}{n}}+d=\dfrac{z^n}{2} \\ x^{\frac{k}{n}}+2d=y^{\frac{m}{n}} \end{cases} \ , \quad d \text{ is the common difference.}$$

$x^{\frac{k}{n}}$, $\dfrac{z^n}{2}$ and $y^{\frac{m}{n}}$ are the arithmetic progression.

Then let $z=2, d=x^{\frac{k}{n}}, m=n+1, k=n+3, (n=1,2,3,\dots,\infty)$,

That it means:
$$\begin{cases} x^{\frac{n+3}{n}}+x^{\frac{n+3}{n}}=\dfrac{2^n}{2} \\ x^{\frac{n+3}{n}}+2x^{\frac{n+3}{n}}=y^{\frac{n+1}{n}} \end{cases} \ .$$

Because $x^{\frac{n+3}{n}}+x^{\frac{n+3}{n}}=\dfrac{2^n}{2}$, in which $4x^{\frac{n+3}{n}}=2^n$, and $x=2^{(\frac{n}{n+3})\log_{(2)}(\frac{1}{4})+\frac{n^2}{n+3}}$

.

Also because $x^{\frac{n+3}{n}}+2x^{\frac{n+3}{n}}=y^{\frac{n+1}{n}}$, in which $3x^{\frac{n+3}{n}}=y^{\frac{n+1}{n}}$,

and $y=2^{(\frac{n}{n+1})\log_{(2)}(\frac{3}{4})+\frac{n^2}{n+1}}$.

Now, just let the base number 2 to be any prime number p ,

These have: $x = p^{\left(\frac{n}{n+3}\right)\log_{(p)}\left(\frac{1}{4}\right)+\frac{n^2}{n+3}}$, $y = p^{\left(\frac{n}{n+1}\right)\log_{(p)}\left(\frac{3}{4}\right)+\frac{n^2}{n+1}}$, $z = p$.

So when $x^{\frac{k}{n}} + y^{\frac{m}{n}} = z^n$, $k>0, m>0, n>0, x>0, y>0, z>0$,

$$m = n+1, k = n+3, (n = 1,2,3,\ldots,\infty) ,$$

$$x = p^{\left(\frac{n}{n+3}\right)\log_{(p)}\left(\frac{1}{4}\right)+\frac{n^2}{n+3}} , \quad y = p^{\left(\frac{n}{n+1}\right)\log_{(p)}\left(\frac{3}{4}\right)+\frac{n^2}{n+1}} , \quad z = p ,$$

and p is any prime number, I call it X-F Theorem 19.

A series of X-F Theorem 19

1) $a x^{\frac{k}{n}} + y^{\frac{m}{n}} = z^n$, $a>0, k>0, m>0, n>0, x>0, y>0, z>0$,

When $a = n+5, m = n+1, k = n+3, (n = 1,2,3,\ldots,\infty)$,

These have: $x = p^{\left(\frac{n}{n+3}\right)\log_{(p)}\left(\frac{1}{4n+20}\right)+\frac{n^2}{n+3}}$, $y = p^{\left(\frac{n}{n+1}\right)\log_{(p)}\left(\frac{3}{4}\right)+\frac{n^2}{n+1}}$, $z = p$,

and p is any prime number.

2) $x^{\frac{k}{n}} + b y^{\frac{m}{n}} = z^n$, $b>0, k>0, m>0, n>0, x>0, y>0, z>0$,

When $b = n+6, m = n+1, k = n+3, (n = 1,2,3,\ldots,\infty)$,

These have: $x = p^{\left(\frac{n}{n+3}\right)\log_{(p)}\left(\frac{1}{4}\right)+\frac{n^2}{n+3}}$, $y = p^{\left(\frac{n}{n+1}\right)\log_{(p)}\left(\frac{3}{4p+24}\right)+\frac{n^2}{n+1}}$, $z = p$,

and p is any prime number.

3) $\quad x^{\frac{k}{n}}+y^{\frac{m}{n}}=c\,z^{n}$, $\quad c>0,k>0,m>0,n>0,x>0,y>0,z>0$,

When $\quad c=n+7,m=n+1,k=n+3,\left(n=1,2,3,...,\infty\right)$,

These have: $\quad x=p^{\left(\frac{n}{n+3}\right)\log_{(p)}\left(\frac{n+7}{4}\right)+\frac{n^2}{n+3}}$, $\quad y=p^{\left(\frac{n}{n+1}\right)\log_{(p)}\left(\frac{3\,p+21}{4}\right)+\frac{n^2}{n+1}}$, $\quad z=p$,

and p is any prime number.

4) $\quad a\,x^{\frac{k}{n}}+b\,y^{\frac{m}{n}}=z^{n}$, $\quad a>0,b>0,k>0,m>0,n>0,x>0,y>0,z>0$,

When $\quad a=n+5,b=n+6,m=n+1,k=n+3,\left(n=1,2,3,...,\infty\right)$,

These have: $\quad x=p^{\left(\frac{n}{n+3}\right)\log_{(p)}\left(\frac{1}{4\,n+20}\right)+\frac{n^2}{n+3}}$, $\quad y=p^{\left(\frac{n}{n+1}\right)\log_{(p)}\left(\frac{3}{4\,p+24}\right)+\frac{n^2}{n+1}}$, $\quad z=p$,

and p is any prime number.

5) $\quad a\,x^{\frac{k}{n}}+y^{\frac{m}{n}}=c\,z^{n}$, $\quad a>0,c>0,k>0,m>0,n>0,x>0,y>0,z>0$,

When $\quad a=n+5,c=n+7,m=n+1,k=n+3,\left(n=1,2,3,...,\infty\right)$,

These have: $\quad x=p^{\left(\frac{n}{n+3}\right)\log_{(p)}\left(\frac{n+7}{4\,n+20}\right)+\frac{n^2}{n+3}}$, $\quad y=p^{\left(\frac{n}{n+1}\right)\log_{(p)}\left(\frac{3\,p+21}{4}\right)+\frac{n^2}{n+1}}$, $\quad z=p$,

and p is any prime number.

6) $\quad x^{\frac{k}{n}}+b\,y^{\frac{m}{n}}=c\,z^{n}$, $\quad b>0,c>0,k>0,m>0,n>0,x>0,y>0,z>0$,

When $b=n+6, c=n+7, m=n+1, k=n+3, (n=1,2,3,\ldots,\infty)$,

These have: $x=p^{(\frac{n}{n+3})\log_{(p)}(\frac{n+7}{4})+\frac{n^2}{n+3}}$, $y=p^{(\frac{n}{n+1})\log_{(p)}(\frac{3p+21}{4p+24})+\frac{n^2}{n+1}}$, $z=p$,

and p is any prime number.

7) $a x^{\frac{k}{n}}+b y^{\frac{m}{n}}=c z^n$, $a>0, b>0, c>0, k>0, m>0, n>0, x>0, y>0, z>0$,

When $a=n+5, b=n+6, c=n+7, m=n+1, k=n+3, (n=1,2,3,\ldots,\infty)$,

These have: $x=p^{(\frac{n}{n+3})\log_{(p)}(\frac{n+7}{4n+20})+\frac{n^2}{n+3}}$, $y=p^{(\frac{n}{n+1})\log_{(p)}(\frac{3p+21}{4p+24})+\frac{n^2}{n+1}}$, $z=p$,

and p is any prime number.

20:

$x^{\frac{k}{m}}+y^{\frac{e}{f}}=z^n$, $k>0, m>0, e>0, f>0, n>0, x>0, y>0, z>0$,

Let $z^n=2p$, it has:
$$\begin{cases} x^{\frac{k}{m}}+d=\dfrac{z^n}{2} \\ x^{\frac{k}{m}}+2d=y^{\frac{e}{f}} \end{cases}, \quad d \text{ is the common difference.}$$

$x^{\frac{k}{m}}$, $\dfrac{z^n}{2}$ and $y^{\frac{e}{f}}$ are the arithmetic progression.

Then let $z=2, k=2n+1, m=3n+1, e=4n+1, f=5n+1, (n=1,2,3,\ldots,\infty)$,

That it means:
$$\begin{cases} x^{\frac{2n+1}{3n+1}} + x^{\frac{2n+1}{3n+1}} = \dfrac{2^n}{2} \\ x^{\frac{2n+1}{3n+1}} + 2x^{\frac{2n+1}{3n+1}} = y^{\frac{4n+1}{5n+1}} \end{cases}.$$

Because $x^{\frac{2n+1}{3n+1}} + x^{\frac{2n+1}{3n+1}} = \dfrac{2^n}{2}$, in which $4x^{\frac{2n+1}{3n+1}} = 2^n$,

and $x = 2^{\left(\frac{3n+1}{2n+1}\right)\log_{(2)}\left(\frac{1}{4}\right) + \frac{n(3n+1)}{2n+1}}$.

Also because $x^{\frac{2n+1}{3n+1}} + 2x^{\frac{2n+1}{3n+1}} = y^{\frac{4n+1}{5n+1}}$, in which $3x^{\frac{2n+1}{3n+1}} = y^{\frac{4n+1}{5n+1}}$,

and $y = 2^{\left(\frac{5n+1}{4n+1}\right)\log_{(2)}\left(\frac{3}{4}\right) + \frac{n(5n+1)}{4n+1}}$.

Now, just let the base number 2 to be any prime number p ,

These have: $x = p^{\left(\frac{3n+1}{2n+1}\right)\log_{(p)}\left(\frac{1}{4}\right) + \frac{n(3n+1)}{2n+1}}$, $y = p^{\left(\frac{5n+1}{4n+1}\right)\log_{(p)}\left(\frac{3}{4}\right) + \frac{n(5n+1)}{4n+1}}$, $z = p$.

So when $x^{\frac{k}{m}} + y^{\frac{e}{f}} = z^n$, $k>0, m>0, e>0, f>0, n>0, x>0, y>0, z>0$,

$k = 2n+1, m = 3n+1, e = 4n+1, f = 5n+1, (n=1,2,3,...,\infty)$,

$x = p^{\left(\frac{3n+1}{2n+1}\right)\log_{(p)}\left(\frac{1}{4}\right) + \frac{n(3n+1)}{2n+1}}$, $y = p^{\left(\frac{5n+1}{4n+1}\right)\log_{(p)}\left(\frac{3}{4}\right) + \frac{n(5n+1)}{4n+1}}$, $z = p$,

and p is any prime number, I call it X-F Theorem 20.

A series of X-F Theorem 20

1) $ax^{\frac{k}{m}} + y^{\frac{e}{f}} = z^n$, $a>0, k>0, m>0, e>0, f>0, n>0, x>0, y>0, z>0$,

When $a=n+5, k=2n+1, m=3n+1, e=4n+1, f=5n+1, (n=1,2,3,\dots,\infty)$,

These have: $x=p^{(\frac{3n+1}{2n+1})\log_{(p)}(\frac{1}{4n+20})+\frac{n(3n+1)}{2n+1}}$, $y=p^{(\frac{5n+1}{4n+1})\log_{(p)}(\frac{3}{4})+\frac{n(5n+1)}{4n+1}}$,

$z=p$, and p is any prime number.

2) $x^{\frac{k}{m}}+b\,y^{\frac{e}{f}}=z^{n}$, $b>0, k>0, m>0, e>0, f>0, n>0, x>0, y>0, z>0$,

When $b=n+4, k=2n+1, m=3n+1, e=4n+1, f=5n+1, (n=1,2,3,\dots,\infty)$,

These have: $x=p^{(\frac{3n+1}{2n+1})\log_{(p)}(\frac{1}{4})+\frac{n(3n+1)}{2n+1}}$, $y=p^{(\frac{5n+1}{4n+1})\log_{(p)}(\frac{3}{4n+16})+\frac{n(5n+1)}{4n+1}}$,

$z=p$, and p is any prime number.

3) $x^{\frac{k}{m}}+y^{\frac{e}{f}}=c\,z^{n}$, $c>0, k>0, m>0, e>0, f>0, n>0, x>0, y>0, z>0$,

When $c=n+3, k=2n+1, m=3n+1, e=4n+1, f=5n+1, (n=1,2,3,\dots,\infty)$,

These have: $x=p^{(\frac{3n+1}{2n+1})\log_{(p)}(\frac{n+3}{4})+\frac{n(3n+1)}{2n+1}}$, $y=p^{(\frac{5n+1}{4n+1})\log_{(p)}(\frac{3n+9}{4})+\frac{n(5n+1)}{4n+1}}$,

$z=p$, and p is any prime number.

4) $a\,x^{\frac{k}{m}}+b\,y^{\frac{e}{f}}=z^{n}$, $a>0, b>0, k>0, m>0, e>0, f>0, n>0, x>0, y>0, z>0$,

When $a=n+5, b=n+4, k=2n+1, m=3n+1, e=4n+1, f=5n+1, (n=1,2,3,\dots,\infty)$,

These have: $x = p^{(\frac{3n+1}{2n+1})\log_{(p)}(\frac{1}{4n+20})+\frac{n(3n+1)}{2n+1}}$,

$y = p^{(\frac{5n+1}{4n+1})\log_{(p)}(\frac{3}{4n+16})+\frac{n(5n+1)}{4n+1}}$,

$z = p$, and p is any prime number.

5) $a x^{\frac{k}{m}} + y^{\frac{e}{f}} = c z^n$, $a>0, c>0, k>0, m>0, e>0, f>0, n>0, x>0, y>0, z>0$,

When $a = n+5, c = n+3, k = 2n+1, m = 3n+1, e = 4n+1, f = 5n+1, (n=1,2,3,...,\infty)$,

These have: $x = p^{(\frac{3n+1}{2n+1})\log_{(p)}(\frac{n+3}{4n+20})+\frac{n(3n+1)}{2n+1}}$, $y = p^{(\frac{5n+1}{4n+1})\log_{(p)}(\frac{3n+9}{4})+\frac{n(5n+1)}{4n+1}}$,

$z = p$, and p is any prime number.

6) $x^{\frac{k}{m}} + b y^{\frac{e}{f}} = c z^n$, $b>0, c>0, k>0, m>0, e>0, f>0, n>0, x>0, y>0, z>0$,

When $b = n+4, c = n+3, k = 2n+1, m = 3n+1, e = 4n+1, f = 5n+1, (n=1,2,3,...,\infty)$,

These have: $x = p^{(\frac{3n+1}{2n+1})\log_{(p)}(\frac{n+3}{4})+\frac{n(3n+1)}{2n+1}}$, $y = p^{(\frac{5n+1}{4n+1})\log_{(p)}(\frac{3n+9}{4n+16})+\frac{n(5n+1)}{4n+1}}$,

$z = p$, and p is any prime number.

7) $a x^{\frac{k}{m}} + b y^{\frac{e}{f}} = c z^n$, $a>0, b>0, c>0, k>0, m>0, e>0, f>0, n>0, x>0, y>0, z>0$,

When $a = n+5, b = n+4, c = n+3, k = 2n+1, m = 3n+1, e = 4n+1, f = 5n+1, (n=1,2,3,...,\infty)$,

These have: $x = p^{(\frac{3n+1}{2n+1})\log_{(p)}(\frac{n+3}{4n+20})+\frac{n(3n+1)}{2n+1}}$, $y = p^{(\frac{5n+1}{4n+1})\log_{(p)}(\frac{3n+9}{4n+16})+\frac{n(5n+1)}{4n+1}}$,

$z=p$, and p is any prime number.

21:

$$x^k + y^m = z^{\frac{e}{n}} \;,\quad k>0, m>0, e>0, n>0, x>0, y>0, z>0 \;,$$

Let $z^{\frac{e}{n}} = 2p$, it has: $\begin{cases} x^k + d = \dfrac{z^{\frac{e}{n}}}{2} \;,\quad d \text{ is any prime number.} \\ x^k + 2d = y^m \end{cases}$

x^k , $\dfrac{z^{\frac{e}{n}}}{2}$ and y^m are the arithmetic progression.

Then let $z=2, d=x^k, k=2n+3, m=3n+4, e=4n+5, (n=1,2,3,...,\infty)$,

That it means: $\begin{cases} x^{2n+3} + x^{2n+3} = \dfrac{2^{\frac{4n+5}{n}}}{2} \\ x^{2n+3} + 2x^{2n+3} = y^{3n+4} \end{cases}$.

Because $x^{2n+3} + x^{2n+3} = \dfrac{2^{\frac{4n+5}{n}}}{2}$, in which $4x^{2n+3} = 2^{\frac{4n+5}{n}}$,

and $x = 2^{(\frac{1}{2n+3})\log_{(2)}(\frac{1}{4}) + \frac{4n+5}{n(2n+3)}}$.

Also because $x^{2n+3} + 2x^{2n+3} = y^{3n+4}$, in which $3x^{2n+3} = y^{3n+4}$,

and $y = 2^{(\frac{1}{3n+4})\log_{(2)}(\frac{3}{4}) + \frac{4n+5}{n(3n+4)}}$.

Now, just let the base number 2 to be any prime number p ,

These have: $x=p^{(\frac{1}{2n+3})\log_{(p)}(\frac{1}{4})+\frac{4n+5}{n(2n+3)}}$, $y=p^{(\frac{1}{3n+4})\log_{(p)}(\frac{3}{4})+\frac{4n+5}{n(3n+4)}}$,

$$z=p .$$

So when $x^k+y^m=z^{\frac{e}{n}}$, $k>0,m>0,e>0,n>0,x>0,y>0,z>0$,

$k=2n+3,m=3n+4,e=4n+5,(n=1,2,3,...,\infty)$,

$x=p^{(\frac{1}{2n+3})\log_{(p)}(\frac{1}{4})+\frac{4n+5}{n(2n+3)}}$, $y=p^{(\frac{1}{3n+4})\log_{(p)}(\frac{3}{4})+\frac{4n+5}{n(3n+4)}}$, $z=p$,

and p is any prime number, I call it X-F Theorem 21.

A series of X-F Theorem 21

1) $a x^k+y^m=z^{\frac{e}{n}}$, $a>0,k>0,m>0,e>0,n>0,x>0,y>0,z>0$,

When $a=n+2,k=2n+3,m=3n+4,e=4n+5,(n=1,2,3,...,\infty)$,

These have: $x=p^{(\frac{1}{2n+3})\log_{(p)}(\frac{1}{4n+8})+\frac{4n+5}{n(2n+3)}}$, $y=p^{(\frac{1}{3n+4})\log_{(p)}(\frac{3}{4})+\frac{4n+5}{n(3n+4)}}$,

$z=p$, and p is any prime number.

2) $x^k+b y^m=z^{\frac{e}{n}}$, $b>0,k>0,m>0,e>0,n>0,x>0,y>0,z>0$,

When $b=n+1,k=2n+3,m=3n+4,e=4n+5,(n=1,2,3,...,\infty)$,

These have: $x=p^{(\frac{1}{2n+3})\log_{(p)}(\frac{1}{4})+\frac{4n+5}{n(2n+3)}}$, $y=p^{(\frac{1}{3n+4})\log_{(p)}(\frac{3}{4n+4})+\frac{4n+5}{n(3n+4)}}$,

$z=p$, and p is any prime number.

3) $\quad x^k + y^m = c\,z^{\frac{e}{n}}$, $\quad c>0, k>0, m>0, e>0, n>0, x>0, y>0, z>0$,

when $\quad c = n+5, k = 2n+3, m = 3n+4, e = 4n+5, (n=1,2,3,\ldots,\infty)$,

These have: $\quad x = p^{(\frac{1}{2n+3})\log_{(p)}(\frac{n+5}{4}) + \frac{4n+5}{n(2n+3)}}$, $\quad y = p^{(\frac{1}{3n+4})\log_{(p)}(\frac{3n+15}{4}) + \frac{4n+5}{n(3n+4)}}$,

$\quad z = p$, and $\quad p$ is any prime number.

4) $\quad a\,x^k + b\,y^m = z^{\frac{e}{n}}$, $\quad a>0, b>0, k>0, m>0, e>0, n>0, x>0, y>0, z>0$,

When $\quad a = n+2, b = n+1, k = 2n+3, m = 3n+4, e = 4n+5, (n=1,2,3,\ldots,\infty)$,

These have: $\quad x = p^{(\frac{1}{2n+3})\log_{(p)}(\frac{1}{4n+8}) + \frac{4n+5}{n(2n+3)}}$, $\quad y = p^{(\frac{1}{3n+4})\log_{(p)}(\frac{3}{4n+4}) + \frac{4n+5}{n(3n+4)}}$,

$\quad z = p$, and $\quad p$ is any prime number.

5) $\quad a\,x^k + y^m = c\,z^{\frac{e}{n}}$, $\quad a>0, c>0, k>0, m>0, e>0, n>0, x>0, y>0, z>0$,

When $\quad a = n+2, c = n+5, k = 2n+3, m = 3n+4, e = 4n+5, (n=1,2,3,\ldots,\infty)$,

These have: $\quad x = p^{(\frac{1}{2n+3})\log_{(p)}(\frac{n+5}{4n+8}) + \frac{4n+5}{n(2n+3)}}$, $\quad y = p^{(\frac{1}{3n+4})\log_{(p)}(\frac{3n+15}{4}) + \frac{4n+5}{n(3n+4)}}$

,

$\quad z = p$, and $\quad p$ is any prime number.

6) $\quad x^k + b\,y^m = c\,z^{\frac{e}{n}}$, $\quad b>0, c>0, k>0, m>0, e>0, n>0, x>0, y>0, z>0$,

When $b=n+1, c=n+5, k=2n+3, m=3n+4, e=4n+5, (n=1,2,3,\ldots,\infty)$,

These have: $x=p^{(\frac{1}{2n+3})\log_{(p)}(\frac{n+5}{4})+\frac{4n+5}{n(2n+3)}}$, $y=p^{(\frac{1}{3n+4})\log_{(p)}(\frac{3n+15}{4n+4})+\frac{4n+5}{n(3n+4)}}$,

$z=p$, and p is any prime number.

7) $a x^{k}+b y^{m}=c z^{\frac{e}{n}}$, $a>0, b>0, c>0, k>0, m>0, e>0, n>0, x>0, y>0, z>0$,

When $a=n+2, b=n+1, c=n+5, k=2n+3, m=3n+4, e=4n+5, (n=1,2,3,\ldots,\infty)$,

These have: $x=p^{(\frac{1}{2n+3})\log_{(p)}(\frac{n+5}{4n+8})+\frac{4n+5}{n(2n+3)}}$, $y=p^{(\frac{1}{3n+4})\log_{(p)}(\frac{3n+15}{4n+4})+\frac{4n+5}{n(3n+4)}}$,

$z=p$, and p is any prime number.

22:

$x^{m}+y^{m}=z^{\frac{k}{n}}$, $m>0, k>0, n>0, x>0, y>0, z>0$,

Let $z^{\frac{k}{n}}=2p$, it has: $\begin{cases} x^{m}+d=\dfrac{z^{\frac{k}{n}}}{2} \\ x^{m}+2d=y^{m} \end{cases}$, d is any prime number.

x^{m} , $\dfrac{z^{\frac{k}{n}}}{2}$ and y^{m} are the arithmetic progression.

Then let $z=2, d=x^{m}, n=2m+3, k=m+2, (m=1,2,3,\ldots,\infty)$,

That it means: $\begin{cases} x^m + x^m = \dfrac{2^{\frac{m+2}{2m+3}}}{2} \\ x^m + 2x^m = y^m \end{cases}$.

Because $x^m + x^m = \dfrac{2^{\frac{m+2}{2m+3}}}{2}$, in which $4x^m = 2^{\frac{m+2}{2m+3}}$,

and $x = 2^{(\frac{1}{m})\log_{(2)}(\frac{1}{4}) + \frac{m+2}{m(2m+3)}}$,

Also because $x^m + 2x^m = y^m$, in which $3x^m = y^m$,

and $y = 2^{(\frac{1}{m})\log_{(2)}(\frac{3}{4}) + \frac{m+2}{m(2m+3)}}$.

Now, just let the base number 2 to be any prime number p ,

These have: $x = p^{(\frac{1}{m})\log_{(p)}(\frac{1}{4}) + \frac{m+2}{m(2m+3)}}$, $y = p^{(\frac{1}{m})\log_{(p)}(\frac{3}{4}) + \frac{m+2}{m(2m+3)}}$,

$z = p$.

So when $x^m + y^m = z^{\frac{k}{n}}$, $m>0, k>0, n>0, x>0, y>0, z>0$,

$n = 2m+3, k = m+2, (m = 1,2,3,...,\infty)$,

$x = p^{(\frac{1}{m})\log_{(p)}(\frac{1}{4}) + \frac{m+2}{m(2m+3)}}$, $y = p^{(\frac{1}{m})\log_{(p)}(\frac{3}{4}) + \frac{m+2}{m(2m+3)}}$, $z = p$,

and p is any prime number, I call it X-F Theorem 22.

A series of X-F Theorem 22

1) $a\,x^m + y^m = z^{\frac{k}{n}}$, $a>0, m>0, k>0, n>0, x>0, y>0, z>0$,

When $a=3m+2, n=2m+3, k=m+2, (m=1,2,3,...,\infty)$,

These have: $x = p^{(\frac{1}{m})\log_{(p)}(\frac{1}{12m+8})+\frac{m+2}{m(2m+3)}}$, $y = p^{(\frac{1}{m})\log_{(p)}(\frac{3}{4})+\frac{m+2}{m(2m+3)}}$,

$z = p$, and p is any prime number.

2) $x^m + b\,y^m = z^{\frac{k}{n}}$, $b>0, m>0, k>0, n>0, x>0, y>0, z>0$,

When $b=4m+2, n=2m+3, k=m+2, (m=1,2,3,...,\infty)$,

These have: $x = p^{(\frac{1}{m})\log_{(p)}(\frac{1}{4})+\frac{m+2}{m(2m+3)}}$, $y = p^{(\frac{1}{m})\log_{(p)}(\frac{3}{16m+8})+\frac{m+2}{m(2m+3)}}$,

$z = p$, and p is any prime number.

3) $x^m + y^m = c\,z^{\frac{k}{n}}$, $c>0, m>0, k>0, n>0, x>0, y>0, z>0$,

When $c=5m+2, n=2m+3, k=m+2, (m=1,2,3,...,\infty)$,

These have: $x = p^{(\frac{1}{m})\log_{(p)}(\frac{5m+2}{4})+\frac{m+2}{m(2m+3)}}$, $y = p^{(\frac{1}{m})\log_{(p)}(\frac{15m+6}{4})+\frac{m+2}{m(2m+3)}}$,

$z = p$, and p is any prime number.

4) $a\,x^m + b\,y^m = z^{\frac{k}{n}}$, $a>0, b>0, m>0, k>0, n>0, x>0, y>0, z>0$,

When $a=3m+2, b=4m+2, n=2m+3, k=m+2, (m=1,2,3,...,\infty)$,

These have:　$x = p^{(\frac{1}{m})\log_{(p)}(\frac{1}{12m+8})+\frac{m+2}{m(2m+3)}}$, 　$y = p^{(\frac{1}{m})\log_{(p)}(\frac{3}{16m+8})+\frac{m+2}{m(2m+3)}}$,

$z = p$, and　p is any prime number.

5)　$a\,x^m + y^m = c\,z^{\frac{k}{n}}$, 　$a>0, c>0, m>0, k>0, n>0, x>0, y>0, z>0$,

When　$a = 3m+2, c = 5m+2, n = 2m+3, k = m+2, (m = 1,2,3,\ldots,\infty)$,

These have:　$x = p^{(\frac{1}{m})\log_{(p)}(\frac{5m+2}{12m+8})+\frac{m+2}{m(2m+3)}}$, 　$y = p^{(\frac{1}{m})\log_{(p)}(\frac{15m+6}{4})+\frac{m+2}{m(2m+3)}}$,

$z = p$, and　p is any prime number.

6)　$x^m + b\,y^m = c\,z^{\frac{k}{n}}$, 　$b>0, c>0, m>0, k>0, n>0, x>0, y>0, z>0$,

When　$b = 4m+2, c = 5m+2, n = 2m+3, k = m+2, (m = 1,2,3,\ldots,\infty)$,

These have:　$x = p^{(\frac{1}{m})\log_{(p)}(\frac{5m+2}{4})+\frac{m+2}{m(2m+3)}}$, 　$y = p^{(\frac{1}{m})\log_{(p)}(\frac{15m+6}{16m+8})+\frac{m+2}{m(2m+3)}}$,

$z = p$, and　p is any prime number.

7)　$a\,x^m + b\,y^m = c\,z^{\frac{k}{n}}$, 　$a>0, b>0, c>0, m>0, k>0, n>0, x>0, y>0, z>0$,

When　$a = 3m+2, b = 4m+2, c = 5m+2, n = 2m+3, k = m+2, (m = 1,2,3,\ldots,\infty)$,

These have:　$x = p^{(\frac{1}{m})\log_{(p)}(\frac{5m+2}{12m+8})+\frac{m+2}{m(2m+3)}}$, 　$y = p^{(\frac{1}{m})\log_{(p)}(\frac{15m+6}{16m+8})+\frac{m+2}{m(2m+3)}}$,

$z = p$, and　p is any prime number.

$$x^{\frac{1}{n}} + y^k = z^{\frac{m}{j}} \quad , \quad n>0, m>0, k>0, x>0, y>0, z>0 \quad ,$$

Let $z^{\frac{m}{j}} = 2p$, it has: $\begin{cases} x^{\frac{1}{n}} + d = \dfrac{z^{\frac{m}{j}}}{2} \\ x^{\frac{1}{n}} + 2d = y^k \end{cases}$, d is any prime number.

$x^{\frac{1}{n}}$, $\dfrac{z^{\frac{m}{j}}}{2}$ and y^k are the arithmetic progression.

Then let $z=2, d=x^{\frac{1}{n}}, n=m+1, k=2m+3, j=3m+4, (m=1,2,3,...,\infty)$,

That it means: $\begin{cases} x^{\frac{1}{m+1}} + x^{\frac{1}{m+1}} = \dfrac{2^{\frac{m}{3m+4}}}{2} \\ x^{\frac{1}{m+1}} + 2x^{\frac{1}{m+1}} = y^{2m+3} \end{cases}$.

because $x^{\frac{1}{m+1}} + x^{\frac{1}{m+1}} = \dfrac{2^{\frac{m}{3m+4}}}{2}$, in which $4x^{\frac{1}{m+1}} = 2^{\frac{m}{3m+4}}$,

and $x = 2^{(m+1)\log_{(2)}(\frac{1}{4}) + \frac{m(m+1)}{3m+4}}$.

Also because $x^{\frac{1}{m+1}} + 2x^{\frac{1}{m+1}} = y^{2m+3}$, in which $3x^{\frac{1}{m+1}} = y^{2m+3}$,

and $y = 2^{(\frac{1}{2m+3})\log_{(2)}(\frac{3}{4}) + \frac{m}{(2m+3)(3m+4)}}$.

Now, just let the base number 2 to be any prime number p ,

These have: $x = p^{(m+1)\log_{(p)}(\frac{1}{4})+\frac{m(m+1)}{3m+4}}$, $y = p^{(\frac{1}{2m+3})\log_{(p)}(\frac{3}{4})+\frac{m}{(2m+3)(3m+4)}}$,

$z = p$.

So when $x^{\frac{1}{n}} + y^k = z^{\frac{m}{j}}$, $n>0, m>0, k>0, x>0, y>0, z>0$,

$n = m+1, k = 2m+3, j = 3m+4, (m=1,2,3,...,\infty)$,

$x = p^{(m+1)\log_{(p)}(\frac{1}{4})+\frac{m(m+1)}{3m+4}}$, $y = p^{(\frac{1}{2m+3})\log_{(p)}(\frac{3}{4})+\frac{m}{(2m+3)(3m+4)}}$, $z = p$,

and p is any prime number, I call it X-F Theorem 23.

A series of X-F Theorem 23

1) $ax^{\frac{1}{n}} + y^k = z^{\frac{m}{j}}$, $a>0, n>0, m>0, k>0, x>0, y>0, z>0$,

When $a = m+4, n = m+1, k = 2m+3, j = 3m+4, (m=1,2,3,...,\infty)$,

These have: $x = p^{(m+1)\log_{(p)}(\frac{1}{4m+16})+\frac{m(m+1)}{3m+4}}$, $y = p^{(\frac{1}{2m+3})\log_{(p)}(\frac{3}{4})+\frac{m}{(2m+3)(3m+4)}}$,

$z = p$, and p is any prime number.

2) $x^{\frac{1}{n}} + by^k = z^{\frac{m}{j}}$, $b>0, n>0, m>0, k>0, x>0, y>0, z>0$,

When $b = m+3, n = m+1, k = 2m+3, j = 3m+4, (m=1,2,3,...,\infty)$,

These have: $x = p^{(m+1)\log_{(p)}(\frac{1}{4})+\frac{m(m+1)}{3m+4}}$, $y = p^{(\frac{1}{2m+3})\log_{(p)}(\frac{3}{4m+12})+\frac{m}{(2m+3)(3m+4)}}$,

$z = p$, and p is any prime number.

3) $x^{\frac{1}{n}}+y^{k}=c\,z^{\frac{m}{j}}$, $c>0,n>0,m>0,k>0,x>0,y>0,z>0$,

When $c=m+2,n=m+1,k=2m+3,j=3m+4,(m=1,2,3,...,\infty)$,

These have: $x=p^{(m+1)\log_{(p)}(\frac{m+2}{4})+\frac{m(m+1)}{3m+4}}$, $y=p^{(\frac{1}{2m+3})\log_{(p)}(\frac{3m+6}{4})+\frac{m}{(2m+3)(3m+4)}}$,

$z=p$, and p is any prime number.

4) $a\,x^{\frac{1}{n}}+b\,y^{k}=z^{\frac{m}{j}}$, $a>0,b>0,n>0,m>0,k>0,x>0,y>0,z>0$,

When $a=m+4,b=m+3,n=m+1,k=2m+3,j=3m+4,(m=1,2,3,...,\infty)$,

These have: $x=p^{(m+1)\log_{(p)}(\frac{1}{4m+16})+\frac{m(m+1)}{3m+4}}$,

$y=p^{(\frac{1}{2m+3})\log_{(p)}(\frac{3}{4m+12})+\frac{m}{(2m+3)(3m+4)}}$, $z=p$,

and p is any prime number.

5) $a\,x^{\frac{1}{n}}+y^{k}=c\,z^{\frac{m}{j}}$, $a>0,c>0,n>0,m>0,k>0,x>0,y>0,z>0$,

When $a=m+4,c=m+2,n=m+1,k=2m+3,j=3m+4,(m=1,2,3,...,\infty)$,

These have: $x=p^{(m+1)\log_{(p)}(\frac{m+2}{4m+16})+\frac{m(m+1)}{3m+4}}$,

$y=p^{(\frac{1}{2m+3})\log_{(p)}(\frac{3m+6}{4})+\frac{m}{(2m+3)(3m+4)}}$, $z=p$,

and p is any prime number.

6) $\quad x^{\frac{1}{n}}+b\,y^k=c\,z^{\frac{m}{j}}$, $\quad b>0,c>0,n>0,m>0,k>0,x>0,y>0,z>0$,

When $\quad b=m+3,c=m+2,n=m+1,k=2m+3,j=3m+4,(m=1,2,3,...,\infty)$,

These have: $\quad x=p^{(m+1)\log_{(p)}(\frac{m+2}{4})+\frac{m(m+1)}{3m+4}}$,

$$y=p^{(\frac{1}{2m+3})\log_{(p)}(\frac{3m+6}{4m+12})+\frac{m}{(2m+3)(3m+4)}}\quad,\quad z=p\ ,$$

and p is any prime number.

7) $\quad a\,x^{\frac{1}{n}}+b\,y^k=c\,z^{\frac{m}{j}}$, $\quad a>0,b>0,c>0,n>0,m>0,k>0,x>0,y>0,z>0$,

When $\quad a=m+4,b=m+3,c=m+2,n=m+1,k=2m+3,j=3m+4,(m=1,2,3,...,\infty)$,

These have: $\quad x=p^{(m+1)\log_{(p)}(\frac{m+2}{4m+16})+\frac{m(m+1)}{3m+4}}$,

$$y=p^{(\frac{1}{2m+3})\log_{(p)}(\frac{3m+6}{4m+12})+\frac{m}{(2m+3)(3m+4)}}\quad,\quad z=p\ ,$$

and p is any prime number.

24:

$$x^k+y^{\frac{m}{j}}=z^{\frac{1}{n}}\ ,\quad n>0,m>0,k>0,x>0,y>0,z>0\ ,$$

Let $z^{\frac{1}{n}}=2\,p$, it has:
$$\begin{cases} x^k+d=\dfrac{z^{\frac{1}{n}}}{2} \\[4mm] x^k+2d=y^{\frac{m}{j}} \end{cases}, \quad d \text{ is the common difference.}$$

x^k , $\dfrac{z^{\frac{1}{n}}}{2}$ and $y^{\frac{m}{j}}$ are the arithmetic progression.

Then let $z=2, d=x^k, n=2m+2, k=m+4, j=m+7, (m=1,2,3,...,\infty)$,

That it means:
$$\begin{cases} x^{m+4}+x^{m+4}=\dfrac{2^{\frac{1}{2m+2}}}{2} \\[4mm] x^{m+4}+2x^{m+4}=y^{\frac{m}{m+7}} \end{cases}.$$

Because $x^{m+4}+x^{m+4}=\dfrac{2^{\frac{1}{2m+2}}}{2}$, in which $4x^{m+4}=2^{\frac{1}{2m+2}}$,

and $x=2^{(\frac{1}{m+4})\log_{(2)}(\frac{1}{4})+\frac{1}{(m+4)(2m+2)}}$.

Also because $x^{m+4}+2x^{m+4}=y^{\frac{m}{m+7}}$, in which $3x^{m+4}=y^{\frac{m}{m+7}}$,

and $y=2^{(\frac{m+7}{m})\log_{(2)}(\frac{3}{4})+\frac{m+7}{m(2m+2)}}$.

Now, just let the base number 2 to be any prime number p ,

These have: $x=p^{(\frac{1}{m+4})\log_{(p)}(\frac{1}{4})+\frac{1}{(m+4)(2m+2)}}$, $y=p^{(\frac{m+7}{m})\log_{(p)}(\frac{3}{4})+\frac{m+7}{m(2m+2)}}$,

$$z=p .$$

So when $x^k+y^{\frac{m}{j}}=z^{\frac{1}{n}}$, $n>0, m>0, k>0, x>0, y>0, z>0$,

$$n=2m+2, k=m+4, j=m+7, (m=1,2,3,\ldots,\infty) \ ,$$

$$x=p^{(\frac{1}{m+4})\log_{(p)}(\frac{1}{4})+\frac{1}{(m+4)(2m+2)}} \ , \quad y=p^{(\frac{m+7}{m})\log_{(p)}(\frac{3}{4})+\frac{m+7}{m(2m+2)}} \ , \quad z=p \ ,$$

and p is any prime number, I call it X-F Theorem 24.

A series of X-F Theorem 24

1) $\quad a x^k + y^{\frac{m}{j}} = z^{\frac{1}{n}} \ , \quad a>0, n>0, m>0, k>0, x>0, y>0, z>0 \ ,$

When $\quad a=3m+1, n=2m+2, k=m+4, j=m+7, (m=1,2,3,\ldots,\infty) \ ,$

These have: $x=p^{(\frac{1}{m+4})\log_{(p)}(\frac{1}{12m+4})+\frac{1}{(m+4)(2m+2)}} \ , \quad y=p^{(\frac{m+7}{m})\log_{(p)}(\frac{3}{4})+\frac{m+7}{m(2m+2)}} \ ,$

$z=p$, and p is any prime number.

2) $\quad x^k + b y^{\frac{m}{j}} = z^{\frac{1}{n}} \ , \quad b>0, n>0, m>0, k>0, x>0, y>0, z>0 \ ,$

When $\quad b=4m+1, n=2m+2, k=m+4, j=m+7, (m=1,2,3,\ldots,\infty) \ ,$

These have: $x=p^{(\frac{1}{m+4})\log_{(p)}(\frac{1}{4})+\frac{1}{(m+4)(2m+2)}} \ , \quad y=p^{(\frac{m+7}{m})\log_{(p)}(\frac{3}{16m+4})+\frac{m+7}{m(2m+2)}} \ ,$

$z=p$, and p is any prime number.

3) $\quad x^k + y^{\frac{m}{j}} = c z^{\frac{1}{n}} \ , \quad c>0, n>0, m>0, k>0, x>0, y>0, z>0 \ ,$

When $\quad c=5m+1, n=2m+2, k=m+4, j=m+7, (m=1,2,3,\ldots,\infty) \ ,$

These have: $x = p^{(\frac{1}{m+4})\log_{(p)}(\frac{5m+1}{4}) + \frac{1}{(m+4)(2m+2)}}$,

$y = p^{(\frac{m+7}{m})\log_{(p)}(\frac{15m+3}{4}) + \frac{m+7}{m(2m+2)}}$, $z = p$,

and p is any prime number.

4) $a x^k + b y^{\frac{m}{j}} = z^{\frac{1}{n}}$, $a>0, b>0, n>0, m>0, k>0, x>0, y>0, z>0$,

When $a = 3m+1, b = 4m+1, n = 2m+2, k = m+4, j = m+7, (m=1,2,3,...,\infty)$,

These have: $x = p^{(\frac{1}{m+4})\log_{(p)}(\frac{1}{12m+4}) + \frac{1}{(m+4)(2m+2)}}$,

$y = p^{(\frac{m+7}{m})\log_{(p)}(\frac{3}{16m+4}) + \frac{m+7}{m(2m+2)}}$, $z = p$,

p is any prime number.

5) $a x^k + y^{\frac{m}{j}} = c z^{\frac{1}{n}}$, $a>0, c>0, n>0, m>0, k>0, x>0, y>0, z>0$,

When $a = 3m+1, c = 5m+1, n = 2m+2, k = m+4, j = m+7, (m=1,2,3,...,\infty)$,

These have: $x = p^{(\frac{1}{m+4})\log_{(p)}(\frac{5m+1}{12m+4}) + \frac{1}{(m+4)(2m+2)}}$,

$y = p^{(\frac{m+7}{m})\log_{(p)}(\frac{15m+3}{4}) + \frac{m+7}{m(2m+2)}}$, $z = p$,

and p is any prime number.

6)　$x^k + b\,y^{\frac{m}{j}} = c\,z^{\frac{1}{n}}$,　$b>0, c>0, n>0, m>0, k>0, x>0, y>0, z>0$,

When　$b=4m+1, c=5m+1, n=2m+2, k=m+4, j=m+7, (m=1,2,3,\ldots,\infty)$,

These have:　$x = p^{(\frac{1}{m+4})\log_{(p)}(\frac{5m+1}{4}) + \frac{1}{(m+4)(2m+2)}}$,

$y = p^{(\frac{m+7}{m})\log_{(p)}(\frac{15m+3}{16m+4}) + \frac{m+7}{m(2m+2)}}$,　$z = p$,

and　p　is any prime number.

7)　$a\,x^k + b\,y^{\frac{m}{j}} = c\,z^{\frac{1}{n}}$,　$a>0, b>0, c>0, n>0, m>0, k>0, x>0, y>0, z>0$,

When　$a=3m+1, b=4m+1, c=5m+1, n=2m+2, k=m+4, j=m+7, (m=1,2,3,\ldots,\infty)$,

These have:　$x = p^{(\frac{1}{m+4})\log_{(p)}(\frac{5m+1}{12m+4}) + \frac{1}{(m+4)(2m+2)}}$,

$y = p^{(\frac{m+7}{m})\log_{(p)}(\frac{15m+3}{16m+4}) + \frac{m+7}{m(2m+2)}}$,　$z = p$,

and　p　is any prime number.

25:

$x^{\frac{1}{n}} + y^{\frac{m}{j}} = z^k$,　$n>0, m>0, k>0, x>0, y>0, z>0$,

Let　$z^k = 2p$, it has:　$\begin{cases} x^{\frac{1}{n}} + d = \dfrac{z^k}{2} \\[2mm] x^{\frac{1}{n}} + 2d = y^{\frac{m}{j}} \end{cases}$,　d　is any prime number.

$x^{\frac{1}{n}}$, $\dfrac{z^k}{2}$ and $y^{\frac{m}{j}}$ are the arithmetic progression.

Then let $z=2, d=x^{\frac{1}{n}}, m=n+7, j=n+8, k=n+9, (n=1,2,3,\dots,\infty)$,

That it means: $\begin{cases} x^{\frac{1}{n}}+x^{\frac{1}{n}}=\dfrac{2^{n+9}}{2} \\ x^{\frac{1}{n}}+2x^{\frac{1}{n}}=y^{\frac{n+7}{n+8}} \end{cases}$.

Because $x^{\frac{1}{n}}+x^{\frac{1}{n}}=\dfrac{2^{n+9}}{2}$, in which $4x^{\frac{1}{n}}=2^{n+9}$,

and $x=2^{(n)\log_{(2)}(\frac{1}{4})+n(n+9)}$.

Also because $x^{\frac{1}{n}}+2x^{\frac{1}{n}}=y^{\frac{n+7}{n+8}}$, in which $3x^{\frac{1}{n}}=y^{\frac{n+7}{n+8}}$,

and $y=2^{(\frac{n+8}{n+7})\log_{(2)}(\frac{3}{4})+\frac{(n+8)(n+9)}{n+7}}$.

Now, just let the base number 2 to be any prime number p ,

These have: $x=p^{(n)\log_{(p)}(\frac{1}{4})+n(n+9)}$,

$y=p^{(\frac{n+8}{n+7})\log_{(p)}(\frac{3}{4})+\frac{(n+8)(n+9)}{n+7}}$, $z=p$.

So when $x^{\frac{1}{n}}+y^{\frac{m}{j}}=z^k$, $n>0, m>0, k>0, x>0, y>0, z>0$,

$m=n+7, j=n+8, k=n+9, (n=1,2,3,\dots,\infty)$,

$x=p^{(n)\log_{(p)}(\frac{1}{4})+n(n+9)}$, $y=p^{(\frac{n+8}{n+7})\log_{(p)}(\frac{3}{4})+\frac{(n+8)(n+9)}{n+7}}$, $z=p$,

and p is any prime number, I call it X-F Theorem 25.

A series of X-F Theorem 25

1) $a x^{\frac{1}{n}} + y^{\frac{m}{j}} = z^k$, $a>0, n>0, m>0, k>0, x>0, y>0, z>0$,

When $a=2n+3, m=n+7, j=n+8, k=n+9, (n=1,2,3,...,\infty)$,

These have: $x = p^{(n)\log_{(p)}\left(\frac{1}{8n+12}\right)+n(n+9)}$,

$y = p^{\left(\frac{n+8}{n+7}\right)\log_{(p)}\left(\frac{3}{4}\right)+\frac{(n+8)(n+9)}{n+7}}$, $z=p$,

and p is any prime number.

2) $x^{\frac{1}{n}} + b y^{\frac{m}{j}} = z^k$, $b>0, n>0, m>0, k>0, x>0, y>0, z>0$,

When $b=2n+4, m=n+7, j=n+8, k=n+9, (n=1,2,3,...,\infty)$,

These have: $x = p^{(n)\log_{(p)}\left(\frac{1}{4}\right)+n(n+9)}$,

$y = p^{\left(\frac{n+8}{n+7}\right)\log_{(p)}\left(\frac{3}{8n+16}\right)+\frac{(n+8)(n+9)}{n+7}}$, $z=p$,

and p is any prime number.

3) $x^{\frac{1}{n}} + y^{\frac{m}{j}} = c z^k$, $c>0, n>0, m>0, k>0, x>0, y>0, z>0$,

When $c=2n+5, m=n+7, j=n+8, k=n+9, (n=1,2,3,...,\infty)$,

These have: $x = p^{(n)\log_{(p)}(\frac{2n+5}{4}) + n(n+9)}$,

$$y = p^{(\frac{n+8}{n+7})\log_{(p)}(\frac{6n+15}{4}) + \frac{(n+8)(n+9)}{n+7}}$$, $z = p$,

and p is any prime number.

4) $\quad a x^{\frac{1}{n}} + b y^{\frac{m}{j}} = z^k$, $a>0, b>0, n>0, m>0, k>0, x>0, y>0, z>0$,

When $a = 2n+3, b = 2n+4, m = n+7, j = n+8, k = n+9, (n = 1,2,3,...,\infty)$,

These have: $x = p^{(n)\log_{(p)}(\frac{1}{8n+12}) + n(n+9)}$,

$$y = p^{(\frac{n+8}{n+7})\log_{(p)}(\frac{3}{8n+16}) + \frac{(n+8)(n+9)}{n+7}}$$, $z = p$,

and p is any prime number.

5) $\quad a x^{\frac{1}{n}} + y^{\frac{m}{j}} = c z^k$, $a>0, c>0, n>0, m>0, k>0, x>0, y>0, z>0$,

When $a = 2n+3, c = 2n+5, m = n+7, j = n+8, k = n+9, (n = 1,2,3,...,\infty)$,

These have: $x = p^{(n)\log_{(p)}(\frac{2n+5}{8n+12}) + n(n+9)}$,

$$y = p^{(\frac{n+8}{n+7})\log_{(p)}(\frac{6n+15}{4}) + \frac{(n+8)(n+9)}{n+7}}$$, $z = p$,

and p is any prime number.

6) $x^{\frac{1}{n}}+b\,y^{\frac{m}{j}}=c\,z^{k}$, $b>0,c>0,n>0,m>0,k>0,x>0,y>0,z>0$,

When $b=2n+4,c=2n+5,m=n+7,j=n+8,k=n+9,(n=1,2,3,...,\infty)$,

These have: $x=p^{(n)\log_{(p)}(\frac{2n+5}{4})+n(n+9)}$,

$y=p^{(\frac{n+8}{n+7})\log_{(p)}(\frac{6n+15}{8n+16})+\frac{(n+8)(n+9)}{n+7}}$, $z=p$,

and p is any prime number.

7) $a\,x^{\frac{1}{n}}+b\,y^{\frac{m}{j}}=c\,z^{k}$, $a>0,b>0,c>0,n>0,m>0,k>0,x>0,y>0,z>0$,

When $a=2n+3,b=2n+4,c=2n+5,m=n+7,j=n+8,k=n+9,(n=1,2,3,...,\infty)$,

These have: $x=p^{(n)\log_{(p)}(\frac{2n+5}{8n+12})+n(n+9)}$,

$y=p^{(\frac{n+8}{n+7})\log_{(p)}(\frac{6n+15}{8n+16})+\frac{(n+8)(n+9)}{n+7}}$, $z=p$,

and p is any prime number.

A series of Anhua-Diophantus Theorem

1) $a\,x^{n}+y^{n}=z^{n}$, $a>0,n>0,x>0,y>0,z>0$,

When $a=n+5,(n=1,2,3,...,\infty)$,

These have: $x=p^{(\frac{1}{n})\log_{(p)}(\frac{1}{4n+20})+1}=p^{\frac{\log_{(p)}(\frac{1}{4n+20})+n}{n}}$,

$$y=p^{(\frac{1}{n})\log_{(p)}(\frac{3}{4})+1}=p^{\frac{\log_{(p)}(\frac{3}{4})+n}{n}} \quad , \quad z=p \quad ,$$

and p is any prime number.

2) $x^n+b\,y^n=z^n$, $b>0,n>0,x>0,y>0,z>0$,

When $b=n+6,(n=1,2,3,...,)$,

These have: $x=p^{(\frac{1}{n})\log_{(p)}(\frac{1}{4})+1}=p^{\frac{\log_{(p)}(\frac{1}{4})+n}{n}}$,

$$y=p^{(\frac{1}{n})\log_{(p)}(\frac{3}{4n+24})+1}=p^{\frac{\log_{(p)}(\frac{3}{4n+24})+n}{n}} \quad , \quad z=p \quad ,$$

and p is any prime number.

3) $x^n+y^n=c\,z^n$, $c>0,n>0,x>0,y>0,z>0$,

When $c=n+7,(n=1,2,3,...,\infty)$,

These have: $x=p^{(\frac{1}{n})\log_{(p)}(\frac{n+7}{4})+1}=p^{\frac{\log_{(p)}(\frac{n+7}{4})+n}{n}}$,

$$y=p^{(\frac{1}{n})\log_{(p)}(\frac{3n+21}{4})+1}=p^{\frac{\log_{(p)}(\frac{3n+21}{4})+n}{n}} \quad , \quad z=p \quad ,$$

and p is any prime number.

4) $a\,x^n+b\,y^n=z^n$, $a>0,b>0,n>0,x>0,y>0,z>0$,

When $a=n+5, b=n+6, (n=1,2,3,...,\infty)$,

These have: $x=p^{(\frac{1}{n})\log_{(p)}(\frac{1}{4n+20})+1} = p^{\frac{\log_{(p)}(\frac{1}{4n+20})+n}{n}}$,

$$y=p^{(\frac{1}{n})\log_{(p)}(\frac{3}{4n+24})+1} = p^{\frac{\log_{(p)}(\frac{3}{4n+24})+n}{n}} \quad , \quad z=p \quad ,$$

and p is any prime number.

5) $a x^n + y^n = c z^n$, $a>0, c>0, n>0, x>0, y>0, z>0$,

When $a=n+5, c=n+7, (n=1,2,3,...,\infty)$,

These have: $x=p^{(\frac{1}{n})\log_{(p)}(\frac{n+7}{4n+20})+1} = p^{\frac{\log_{(p)}(\frac{n+7}{4n+20})+n}{n}}$,

$$y=p^{(\frac{1}{n})\log_{(p)}(\frac{3n+21}{4})+1} = p^{\frac{\log_{(p)}(\frac{3n+21}{4})+n}{n}} \quad , \quad z=p \quad ,$$

and p is any prime number.

6) $x^n + b y^n = c z^n$, $b>0, c>0, n>0, x>0, y>0, z>0$,

When $b=n+6, c=n+7, (n=1,2,3,...,\infty)$,

These have: $x=p^{(\frac{1}{n})\log_{(p)}(\frac{n+7}{4})+1} = p^{\frac{\log_{(p)}(\frac{n+7}{4})+n}{n}}$,

$$y=p^{(\frac{1}{n})\log_{(p)}(\frac{3n+21}{4n+24})+1} = p^{\frac{\log_{(p)}(\frac{3n+21}{4n+24})+n}{n}} \quad , \quad z=p \quad ,$$

and p is any prime number.

7) $ax^n + by^n = cz^n$, $a>0, b>0, c>0, n>0, x>0, y>0, z>0$,

When $a=n+5, b=n+6, c=n+7, (n=1,2,3,\ldots,\infty)$,

These have: $x = p^{(\frac{1}{n})\log_{(p)}(\frac{n+7}{4n+20})+1} = p^{\frac{\log_{(p)}(\frac{n+7}{4n+20})+n}{n}}$,

$y = p^{(\frac{1}{n})\log_{(p)}(\frac{3n+21}{4n+24})+1} = p^{\frac{\log_{(p)}(\frac{3n+21}{4n+24})+n}{n}}$, $z = p$,

and p is any prime number.

A series of Beal Theorem

1) $kA^x + B^y = C^z$, $k>0, x>0, y>0, z>0, A>0, B>0, C>0$,

When $k=2n+1, (n=1,2,3,\ldots,\infty)$,

These have: $A=p^{n+1}, B=p^n, x=n, y=n+1, C=p, z=n(n+1)+\log_{(p)}(2n+2)$,

That it means: $(2n+1)(p^{n+1})^n + (p^n)^{n+1} = p^{n(n+1)+\log_{(p)}(2n+2)}$,

and p is any prime number.

2) $A^x + mB^y = C^z$, $m>0, x>0, y>0, z>0, A>0, B>0, C>0$,

When $m=3n+1, (n=1,2,3,\ldots,\infty)$,

These have: $A=p^{n+1}, B=p^n, x=n, y=n+1, C=p, z=n(n+1)+\log_{(p)}(3n+2)$,

That it means: $\left(p^{n+1}\right)^n+(3n+1)\left(p^n\right)^{n+1}=p^{n(n+1)+\log_{(p)}(3n+2)}$,

and p is any prime number.

3) $A^x+B^y=jC^z$, $j>0, x>0, y>0, z>0, A>0, B>0, C>0$,

When $j=2n, (n=1,2,3,\ldots,\infty)$,

These have: $A=p^{n+1}, B=p^n, x=n, y=n+1, C=p, z=n(n+1)+\log_{(p)}\left(\dfrac{1}{n}\right)$,

That it means: $\left(p^{n+1}\right)^n+\left(p^n\right)^{n+1}=(2n)\,p^{n(n+1)+\log_{(p)}\left(\frac{1}{n}\right)}$,

and p is any prime number.

4) $kA^x+mB^y=C^z$, $k>0, m>0, x>0, y>0, z>0, A>0, B>0, C>0$,

When $k=2n+1, m=3n+1, (n=1,2,3,\ldots,\infty)$,

These have: $A=p^{n+1}, B=p^n, x=n, y=n+1, C=p, z=n(n+1)+\log_{(p)}(5n+2)$,

That it means: $(2n+1)\left(p^{n+1}\right)^n+(3n+1)\left(p^n\right)^{n+1}=p^{n(n+1)+\log_{(p)}(5n+2)}$,

and p is any prime number.

5) $kA^x+B^y=jC^z$, $k>0, j>0, x>0, y>0, z>0, A>0, B>0, C>0$,

When $k=2n+1, j=2n, (n=1,2,3,\ldots,\infty)$,

These have: $A = p^{n+1}, B = p^n, x = n, y = n+1, C = p, z = n(n+1) + \log_{(p)}(\frac{n+1}{n})$,

That it means: $(2n+1)(p^{n+1})^n + (p^n)^{n+1} = (2n) p^{n(n+1) + \log_{(p)}(\frac{n+1}{n})}$,

and p is any prime number.

6) $A^x + m B^y = j C^z$, $m > 0, j > 0, x > 0, y > 0, z > 0, A > 0, B > 0, C > 0$,

When $m = 3n+1, j = 2n, (n = 1, 2, 3, \ldots, \infty)$,

These have: $A = p^{n+1}, B = p^n, x = n, y = n+1, C = p, z = n(n+1) + \log_{(p)}(\frac{3n+2}{2n})$,

That it means: $(p^{n+1})^n + (3n+1)(p^n)^{n+1} = (2n) p^{n(n+1) + \log_{(p)}(\frac{3n+2}{2n})}$,

and p is any prime number.

7) $k A^x + m B^y = j C^z$, $k > 0, m > 0, j > 0, x > 0, y > 0, z > 0, A > 0, B > 0, C > 0$,

When $k = 2n+1, m = 3n+1, j = 2n, (n = 1, 2, 3, \ldots, \infty)$,

These have: $A = p^{n+1}, B = p^n, x = n, y = n+1, C = p, z = n(n+1) + \log_{(p)}(\frac{5n+2}{2n})$,

That it means: $(2n+1)(p^{n+1})^n + (3n+1)(p^n)^{n+1} = (2n) p^{n(n+1) + \log_{(p)}(\frac{5n+2}{2n})}$,

and p is any prime number.

So bored that I feel a little tired, haha...